THE MORPHOLOGY OF GYMNOSPERMS

THE MORPHOLOGY OF GYMNOSPERMS

THE STRUCTURE AND EVOLUTION OF PRIMITIVE SEED-PLANTS

K.R. Sporne
M.A, PH.D, FLS
Fellow of Downing College, Cambridge
and University Lecturer in Botany

United Book Prints

Reprinted in India by:

UNITED BOOK PRINTS
(Imprint of Scientific Publishers, INDIA)
4806/24, Ansari Road, Daryaganj,
New Delhi 110 002, INDIA
E-mail: info@scientificpub.com
Website : http://www.scientificpub.com

Print : 2020

ISBN: 978-93-83692-30-9
eISBN: 978-93-88172-50-9

Printed in India

TO MY WIFE,
HELEN

Contents

Preface

Many textbooks have been published which have included a chapter or two on living gymnosperms, with perhaps a brief mention of some fossil forms. Likewise, there are several textbooks of palaeobotany which describe the numerous fossil members of the group. However, there has been no textbook since 1934[5] that is devoted entirely to the gymnosperms and which gives equal emphasis to both living and fossil members. During this interval a great many remarkable discoveries have been made which have profoundly altered our interpretation of the group. Within the limits of 50,000 words, however, it is difficult to do justice to such a large and fascinating assemblage of plants. So much condensation is necessary that some of the more controversial issues cannot be given the full consideration that they deserve. To compensate for this, there is an extensive bibliography that will enable the reader to follow up those aspects that interest him.

Almost all the illustrations have been re-drawn from published accounts, and the source of each is acknowledged by name. The reader may complain that the figures are sometimes too small or too crowded, but I hope that he will agree that even a small illustration is better than none.

The writer of a book of this kind owes a debt of gratitude to so many people that they cannot all be acknowledged individually—one's teachers, one's colleagues and even one's pupils, all of whom have helped in the formation of one's ideas. I must, however, single out the late Dr Hugh Hamshaw Thomas, who did more than anyone else to arouse and foster an interest in morphology. It was he who first impressed upon me the importance of fossils in the proper understanding of living organisms.

I am very grateful to Professor J. do Amaral Franco, of the Instituto Superior de Agronomia, Lisbon, for checking the nomenclature of the living gymnosperms, and to Dr W. S. Lacey, of the University College of North Wales, Bangor, for his great kindness in reading the final draft. To my wife, to whom I dedicate this book, I am particularly indebted for her constant encouragement, advice, and helpful criticism.

K.R.S.

Cambridge

I

Introduction

> ἔνια δὲ καὶ ἐν λοβῷ, τὰ δ' ἐν ὑμένι, τὰ δ' ἐν ἀγγείῳ, τὰ δὲ καὶ γυμνόσπερμα τελείως.
>
> Some seeds again are enclosed in a pod, some in a husk, some in a vessel, and some are completely naked.

It is not easy to express in simple terms what is implied by the title of this book, for both 'morphology' and 'gymnosperm' are words which have had a long history and whose meanings have undergone some change with the passage of time. It was the German poet and philosopher Goethe who, in 1790, first used the term 'morphology', and it was most appropriate that it should have been compounded of two Greek words, for the ancient Greeks (and, in particular, Aristotle) are regarded as the founders of this branch of natural history. The first known use of the word 'gymnosperm' was, in fact, by one of Aristotle's pupils—Theophrastus—who used it to describe plants whose seeds are unprotected. The quotation at the head of this chapter, containing the word *γυμνόσπερμα*, is taken from his *Enquiry into Plants* (*περί φυτῶν ἱστορία*), one of the earliest known textbooks of botany, written as long ago as 300 B.C.

Literally translated, 'morphology' implies no more than the study of form. However, if this were all, then surely there must have been some knowledge of the subject long before ancient Greek times. Indeed, man's very survival must always have depended upon his ability to remember which plants were edible and which poisonous; for this he had to know something of their form and structure. Thus, morphology in the narrow sense must have been developing gradually over a vast period of time before man became civilised, and the ancient Greeks must have inherited a large corpus of knowledge on which to build. However, till then, such knowledge had been put to purely utilitarian purposes: it now took on a deeper significance. It is in this respect that the Greeks can truly be called the founders of morphology.

On reading the works of Theophrastus, one discovers what an astonishingly clear insight he had into the problems confronting the morphologist. Many of the problems that he discussed then are still being discussed today, e.g. the principles underlying the classification of plants, and the criteria to be used when defining the essential parts of a plant. The opening sentence of *Enquiry into Plants* is particularly interesting. It has been translated by Sir Arthur Hort[12] as follows: 'In considering the distinctive characters of plants and their nature generally one must take into account their parts, their qualities, the ways in which their life originates and the course which it follows in each case.' As Arber[2] has pointed out, such comprehensiveness is remarkable indeed, for here are the first glimmerings of the modern studies, not only of anatomy, but also of physiology and biochemistry, reproduction, ontogeny and life histories. Furthermore, it is clear from his notes on some of the 400 species which he described that he was also aware of the importance of ecology. Aristotelian morphologists had a very broad approach to their subject, and in this respect, therefore, were in advance of many eighteenth-century botanists, whose studies were often much more narrowly restricted. The Greek approach was, in fact, much more like that of modern morphologists, whose aims, according to Arber, should be 'to connect into one coherent whole all that may be held to belong to the intrinsic nature of a living being', of which structure is but one aspect.

During the last hundred years the scope of morphology has been greatly extended by new developments in technique; but, more than this, it has undergone a complete reorientation, with the development of the concept of evolution. The adult form and structure of plants are now seen to be the outcome, not only of a series of developmental processes, but also of a long series of evolutionary changes, extending over hundreds of millions of years. Palaeobotany (the study of fossil plants) is now, therefore, essential to the morphologist who, besides attempting to see living plants in relation to each other, must also try to see them in their relationship to plants long since extinct. He is concerned not only with taxonomy (the science of classification), but also with phylogeny (the study of evolutionary relationships).

To say that the gymnosperms are seed-plants whose seeds are unprotected may mean little to those who are unfamiliar with the morphological nature of the seed or with the full implications of 'protection'. Indeed, it is very difficult to define the group in exact terms, for it contains a number of widely different sub-groups which have been quite distinct for a very long time—at least since Palaeozoic times. In order to understand the seed it is necessary to know something of the life history of other groups of plants and of the place which seed-plants occupy in the classification of the plant kingdom.

On the basis of morphological complexity, four levels of plant organisation are usually recognised: thallophytes, bryophytes, pteridophytes and seed-plants. The first of these, comprising Bacteria, Myxomycetes ('Slime fungi'), Algae (sea-weeds, etc.) and Fungi, are primarily aquatic plants, whose structure ranges from unicellular or filamentous to thalloid. By contrast, the remaining groups (sometimes classed together as the Cormophyta) are more or less adapted to a terrestrial life and show much greater structural complexity, possessing either a rhizoidal part or a root, and (in most cases) an aerial shoot with stem and leaves.

Furthermore, their life cycles are similar, such differences as do occur being no more than variations on a common basic pattern.

This basic life cycle, as exhibited by the bryophytes (mosses and liverworts), is illustrated diagrammatically in Fig. 1. In these plants the gametophyte generation is dominant, the sporophyte remaining attached to it throughout its life, and being dependent upon it to a greater or lesser extent for its nutrition. The gametophyte, as the name suggests, is the generation which bears the sex organs (a flask-shaped

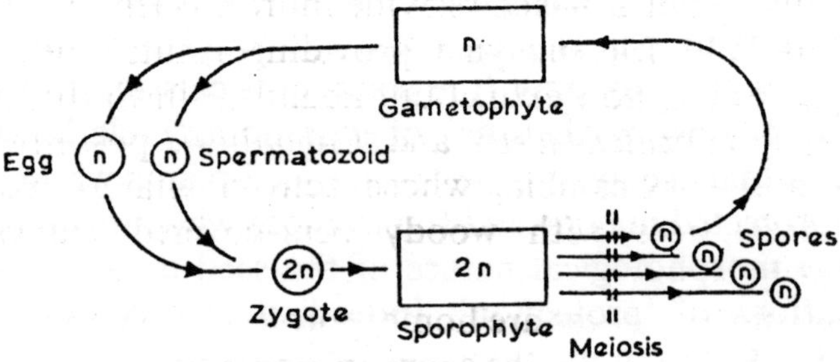

FIG. 1

Basic life cycle of bryophytes and homosporous pteridophytes

archegonium containing a single egg-cell, and a sac-like antheridium from which are released large numbers of flagellated spermatozoids). The spermatozoids require a film of water in which to swim, if they are to reach an archegonium and effect fertilisation. Sexual fusion gives rise to a zygote which, because of its double content of nuclear material, is described as diploid (2n). By a series of mitotic divisions, the zygote develops into the diploid sporophyte generation. Ultimately, this gives rise to non-motile spores which, in most cases, are wind-dispersed. They are formed by a process known as meiosis, involving two nuclear divisions (with the result that four spores are produced from each spore-mother-cell), during which their nuclear content is reduced once more to the haploid condition (n). Under favourable conditions, the spores develop, by mitotic divisions, into haploid gametophytes and the life cycle is completed. Just as the spore is the first cell of the gametophyte generation, so the zygote is the first cell of the sporophyte.

Fig. 1 applies equally well to most pteridophytes, in so far as it is merely a diagrammatic representation. However, in this group (comprising lycopods, horsetails, ferns, etc.) the sporophyte grows much larger than the tiny gametophyte and soon becomes independent of it, developing a stem, leaves and (in most cases) roots. Accompanying this increased size is a greater diversity of tissue types, including xylem and phloem which provide more efficient conducting channels, and sclerenchyma providing greater mechanical strength. Some, now extinct but abundant in Carboniferous times (e.g. *Lepidodendron* and *Calamites*), possessed vascular- and cork-cambia, whose activity enabled them to grow into trees with woody cork-covered trunks and branches.

Most pteridophytes are homosporous and monoecious, as indicated in Fig. 1 (i.e. the spores produced by any particular individual are all of equal size and develop into gametophytes—'prothalli'—bearing both archegonia and antheridia). With such an arrangement, the probability of self-fertilisation is relatively high and evolutionary progress is correspondingly slow. Self-fertilisation is generally regarded as disadvantageous, even in higher plants, but in the case of gametophtyes it is particularly so, since all the sporophytes produced in this way are necessarily homozygous for all characters (a fact which is frequently overlooked). It may be prevented, or made less probable, if the sex organs on a prothallus ripen at different times (e.g. in fern prothalli, the antheridia commonly ripen before the archegonia), or if there is some incompatibility mechanism: but, in any case, cross-fertilisation can only occur if there is sufficient water for the spermatozoids to swim from one prothallus to another. Furthermore, the prothallus lacks a waterproof cuticle, such as covers the aerial parts of the sporophyte, and, for this reason also, it is restricted to moist habitats. Thus, however well adapted the sporophyte may be to a terrestrial existence, it is nevertheless restricted to those habitats where the prothallus can grow and become fertilised. The ecology of homosporous pteridophytes is to a large extent, therefore, that of their prothalli.

Some pteridophytes have escaped from this gametophytic limitation by becoming 'endosporic', i.e. the prothallus is retained within the spore wall, scarcely protruding from it, even at maturity when the spore cracks open. The prothalli of such pteridophytes are invariably dioecious, i.e. the antheridia and archegonia are on different prothalli, thus completely preventing self-fertilisation. Such pteridophytes are also heterosporous, i.e. the spores are of two sizes—'megaspores' and 'microspores'. Although it may not be immediately obvious, this difference in size is concerned with the development of a cross-fertilisation process which is efficient even in relatively dry habitats. The male prothallus within the microspore can be reduced to the minimum size necessary to contain a number of spermatozoids. At the same time, reduction in size allows large numbers to be produced. On the other hand, the female prothallus must contain sufficient food reserves to support the embryo sporophyte until it is large enough to become independent and must, therefore, be relatively large, i.e. a megaspore. Megaspores are too large to be carried very far from the parent plant, but the microspores are light enough to be carried great distances by air currents, and they can be produced in such numbers that some may be expected to land on the female prothallus when it is exposed by the rupture of the spore coat. The distance that the spermatozoids then have to swim in order to reach the archegonia is, therefore, minute and very little moisture suffices. Efficient as this mechanism may be in relatively dry habitats, it is no less efficient in aquatic habitats, for water currents can be as effective as air currents. Thus, heterosporous pteridophytes are able to exploit a much wider range of habitats than are homosporous plants, while at the same time achieving 100 per cent cross-fertilisation.

The number of megaspores in each megasporangium varies considerably throughout the pteridophytes, but in most cases it is four, or a multiple of four. However, there are some (e.g. some species of *Selaginella*) in which some of the four spores in a tetrad are abortive, with the result that there are only two functional megaspores, or even just a

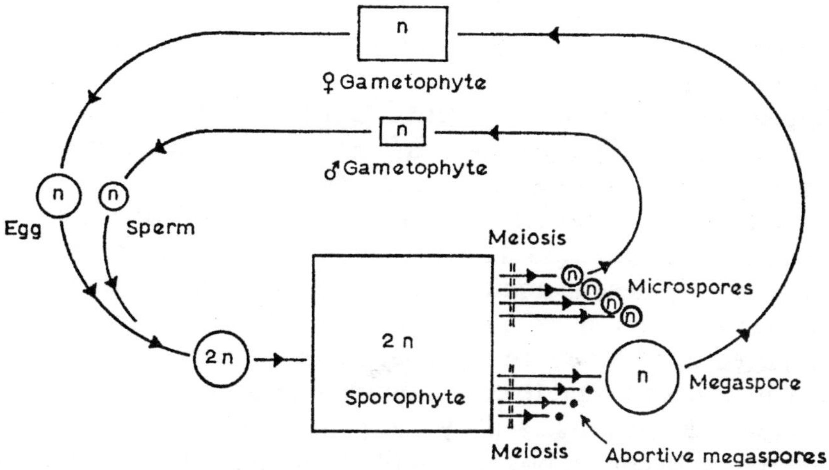

FIG. 2
Life cycle of seed-plants

solitary one, in each sporangium. A single functional megaspore was characteristic also of the fossil genera *Lepidocarpon* and *Miadesmia*. The life cycle of such plants is represented diagrammatically in Fig. 2 (which will readily be seen to be merely a variant of Fig. 1).

We are now in a position to consider seed-plants, for they too have this kind of life cycle. Those species of *Selaginella* which have only one megaspore in each sporangium tend to retain it within the sporangium for a long time; the megaspore may rupture *in situ*, the archegonia in the female prothallus may even become fertilised, and a young sporeling may begin to develop while still on the parent sporophyte. This behaviour forms a close parallel with that of seed-plants. There are two important differences, however. Firstly, in seed-plants the megasporangium does not dehisce, and, secondly, it is almost completely enclosed in one or more envelopes ('integuments'). Up to the stage of fertilisation this compound structure, consisting of sporangium and integument (or integuments), is termed an 'ovule'; after fertilisation it develops into a seed which, at maturity, is shed from the parent plant. An ovule, therefore, may be defined as 'an indehiscent, integumented megasporangium containing a single functional megaspore'. The microspores of seed-plants are usually called 'pollen-grains'.

Seed-plants (Spermaphyta) are usually classified into two major groups, gymnosperms and angiosperms, on the basis of the degree of protection that is afforded to the ovule at the time of pollination. In gymnosperms the ovules are freely exposed, whereas in angiosperms ('flowering plants') they are enclosed in an ovary. Of the two, the angiosperms are believed to be the more highly advanced, since protection of the ovules is thought to confer on them biological advantages which are not possessed by the gymnosperms. The latter, therefore, occupy a position in the plant kingdom which lies between the pteridophytes and the flowering plants. To what extent they represent true evolutionary links between these two groups, however, is not immediately obvious and will be discussed later. In any case, there are many who regard the gymnosperms as representing no more than a particular level of evolution instead of regarding them as a taxonomic group. For this reason, they are referred to throughout this book as gymnosperms, rather than as the Gymnospermae. As long ago as 1948, Arnold[35a] wrote: '. . . it is now believed that the term Gymnospermae as a group name has outlived its usefulness and should be dropped, or it should be retained only as a common name without taxonomic status in modern classification.'

The structure of a gymnosperm seed can best be understood by following the various stages in its development. These are illustrated in Figs 3A–E, for a hypothetical primitive type of seed, as seen in longitudinal section. Some apology should perhaps be made for taking an imaginary example, rather than illustrating a genuine living or fossil ovule, but only by this means is it possible to emphasise the various points of importance. Actual ovules and seeds will, of course, be described in detail in the appropriate chapters. Fig. 3A represents a very young ovule in which the integument (3) is just beginning to grow up round the megasporangium wall (nucellus) (1). Within the nucellus is a single megaspore-mother-cell (2) about to undergo meiosis. In Fig. 3B meiosis has given rise to a single functional megaspore (4) and three abortive ones (5). Fig. 3C illustrates the first stage in the development of the female prothallus (7) within the megaspore-wall (6); as yet, no cross-walls have been laid down between the nuclei (the 'free-nuclear' state). By now the integument has grown up over the nucellus, leaving a small hole ('micropyle') (8) at the apex. In Fig. 3D cell-walls are being laid down between the nuclei of the female prothallus, a process which, in most gymnosperms, continues until the whole of the prothallus becomes cellular. Fig. 3E shows that apical region of this hypothetical ovule at maturity, i.e. at the time of pollination. The integument has three layers, as is commonly the case in gymnosperms—an outer fleshy layer (sarcotesta) (10), a middle stony layer (sclerotesta) (9), and an inner fleshy layer (inner sarcotesta) (11). Pollen-grains (12) have entered the micropyle and are grouped together in the pollen-chamber formed by the breaking down of the apical regions of the nucellus. The megaspore-wall, too, has dissolved away in this region, so that the spermatozoids, when released from the pollen-grains, will have direct access to the archegonia (13) at the apex of the cellular female prothallus (7).

The kind of ovule illustrated here is usually described as having the integument fused with the nucellus at the apex. However, it will readily be appreciated that no actual process

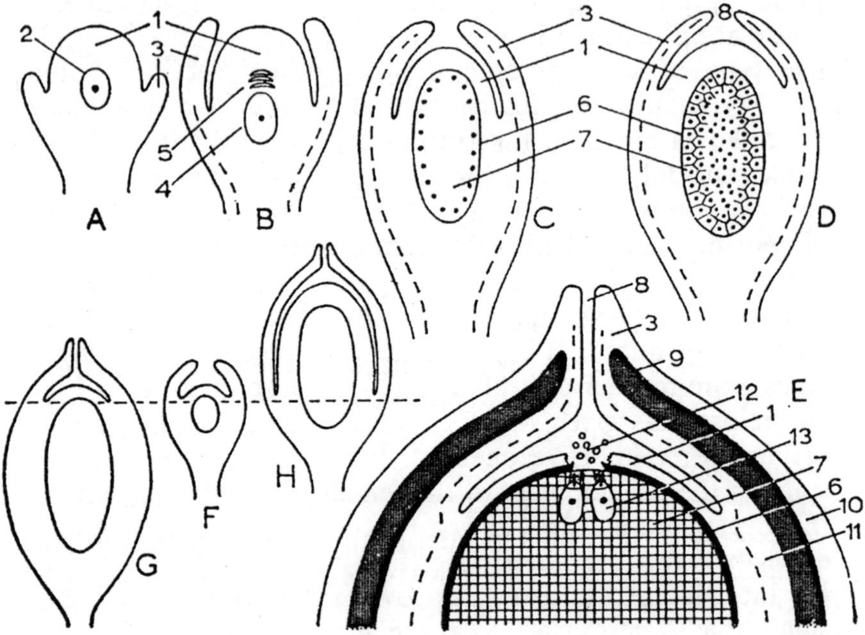

FIG. 3

Development of gymnosperm ovules

A, very young ovule. B, slightly older ovule in which meiosis has occurred. C, ovule with female prothallus at free-nuclear stage. D, ovule with female prothallus becoming cellular. E, apex of hypothetical ovule after pollination.

F, G and H illustrate the way in which differential growth-rates can give rise to different kinds of seed. Extra growth of the chalazal region of a young ovule, F, gives the seed G, in which the integument is 'fused' to the nucellus. Extra growth of the micropylar end gives seed H, in which the integument is free from the nucellus.

(1, nucellus; 2, megaspore-mother-cell; 3, integument; 4, functional megaspore; 5, abortive megaspores; 6, megaspore wall; 7, female prothallus; 8, micropyle; 9, stony layer, or sclerotesta; 10, outer fleshy layer, or sarcotesta; 11, inner fleshy layer; 12, pollen-chamber with pollen-grains; 13, archegonium)

of fusion is involved; rather, it is a form of intercalary growth which is involved, for the basal (or chalazal) end of the ovule grows faster than the micropylar end. By contrast, ovules described as having the integument free from the nucellus are the result of the micropylar end growing faster than the chalazal end. These differences are illustrated in Figs. 3F–H, where G represents the former type in which the chalazal end has enlarged to produce a 'hyposperm', as it is sometimes called, while H represents the latter kind in which the chalazal end has enlarged relatively little.

Until the early part of the present century it was widely believed that, among vascular plants, sporangia must necessarily be borne on leaves, and that such fertile leaves ('sporophylls') are an integral part of their organisation. Since ovules are essentially sporangia, they too must be supposed to be borne on leaves, according to this 'Classical Theory'. This idea grew up, almost unchallenged, out of the suggestion put forward by Goethe,[33] in his *Metamorphosis of Plants* (1790), that all the lateral appendages of flowering plants have the same intrinsic leaf-like nature; leaves, bracts, bracteoles, sepals, petals, stamens and carpels were all to be regarded as manifestations of the same kind of lateral organ. Thus it came about that three organs, and only three, were thought to be essential for the building up of a plant—roots, stems and leaves (some sterile and some fertile). Whatever the truth of this, in so far as flowering plants are concerned, (and there have been many critics in recent years), some gymnosperms, notably Coniferales and Taxales, are particularly difficult to reconcile with it. For many years there was bitter controversy concerning the correct interpretation of the cone-scale, but then the discovery of fossil pteridophytes entirely without leaves demonstrated conclusively that the classical theory had been misleading. It was now possible to accept the idea that, while some plants bear their reproductive organs on leaves, others have them on stems—an idea that applies no less to the gymnosperms than it does to pteridophytes.

Important as reproductive organs are in classification, vegetative organs are also of importance. Among gymnosperms there are two widely different types of leaf, for which the terms 'megaphyll' and 'microphyll' may be borrowed from the pteridophytes. The former is applied to the large branching frond of a fern, while the latter is used to describe the tiny unbranched scale-like leaf of the lycopod. Likewise, in gymnosperms, the term megaphyll may be used to describe relatively large leaves that branch like a fern frond, and have branching veins. The term microphyll may be used for small leaves with one or two veins (or, in some cases, rather larger leaves with parallel venation). In borrowing these terms, it is important to realise that one does not imply any direct phylogenetic descent of microphyllous gymnosperms from microphyllous pteridophytes, for, despite the claims of some morphologists (e.g. Greguss[11]), it is most unlikely that lycopods played any part in the ancestry of conifers. The leaves of the Ginkgoales do not fit well into either category, being either dichotomous or fan-shaped with dichotomous veins; however, recent fossil finds suggest affinities with microphylls, rather than with megaphylls.

There are also two very different kinds of secondary wood to be found in the gymnosperms, known respectively as 'manoxylic' and 'pycnoxylic'. The former is soft and relatively sparse, with very wide parenchymatous rays, while the latter is dense, making up the main bulk of the trunk and branches, with very small wood-rays. Whereas manoxylic wood is useless commercially, pycnoxylic gymnosperms provide a most important part of the world's timber (e.g. deal, pitch pine, etc.). That the distinction between these two types of wood is important taxonomically is clear from the correlation that exists between wood structure, leaf form and seed symmetry. In general it may be said that manoxylic wood is associated with megaphyllous leaves and radial symmetry in the seeds, while pycnoxylic wood is associated with microphyllous leaves and bilateral symmetry in the seeds. Such a correlation would suggest that the gymnosperms represent a diphyletic group, i.e. one derived from two distinct ancestral origins. However, further discussion will be deferred, since any decision must clearly depend on a detailed knowledge of the earliest fossil representatives of the group and of its possible forerunners.

Among pteridophytes there is to be found a very wide range of primary vascular systems, including such types as protosteles, medullated protosteles, solenosteles, dictyosteles, polycyclic steles, polysteles and actinosteles. By contrast, the gymnosperms exhibit a relatively narrow range. While it is true that some are polystelic and that some have co-axial cylinders of secondary wood, formed from anomalous cambia, the vast majority are monostelic. Among these, however, there are some differences in the arrangement of the primary wood, particularly in fossil members of the group.

The most primitive pteridophytes have what is described as a 'solid protostele'. In this there is a single solid rod of xylem tracheids surrounded by phloem and endodermis. All the tissues are primary, in that the cells are laid down in the apical meristem; subsequent developmental changes are limited to change in shape and to differentiation of the cell-wall without any further cell-divisions. Some xylem tracheids become lignified before they have finished elongating, and constitute what is known as the 'protoxylem'. The thickening in such elements is usually annular or helical. By contrast, 'metaxylem' elements lignify later, and may show a variety of patterns of lignification, e.g. scalariform, reticulate, pitted, etc. In some species, or in certain regions of the plant, the protoxylem is at the centre of the xylem rod ('endarch'), the metaxylem developing centrifugally round it. In others the protoxylem lies to the outside of the metaxylem ('exarch'), and the latter develops centripetally. In yet others the protoxylem is 'mesarch', with centrifugal metaxylem to the outside of it and centripetal xylem towards the centre.

It is generally believed that some, at least, of the gymnosperms evolved from very early ferns, or from their ancestors, and that here, too, the solid protostele represents the most primitive type. Fig. 4A illustrates the kind of woody stem that the earliest members of the group might have possessed. In the centre is a solid rod of primary wood with mesarch protoxylems, surrounded by a zone of secondary wood. Figs. 4B–F represent, in a very diagrammatic way, a number of woody stems which are known to have existed in Palaeozoic times, some from the Upper Devonian, some from the Lower Carboniferous and some from the Upper Carboniferous. *Heterangium Grievii* (Fig. 4B) was very similar to the hypothetical ancestral type, except for the parenchyma cells among the metaxylem tracheids. In transverse sections of the stem these give the impression of separating the tracheids into numerous distinct areas, but the three-dimensional arrangement was actually a meshwork of xylem strands whose interstices were filled with parenchyma. Only the outermost strands contained any protoxylem (in a mesarch position), and some of these turned outwards to become leaf-traces, while the rest continued upwards as 'reparatory strands'. The secondary wood was limited in amount and was manoxylic (represented in the diagram by the convention of uneven spacing of the radial lines).

Calamopitys (Fig. 4D) and *Lyginopteris* (Fig. 4F) were also manoxylic. They illustrate the way in which the primary wood tended to become further reduced in amount. In the former there were a few mesarch strands of primary wood surrounding a central region of mixed tracheids and parenchyma, whereas in the latter the central regions were composed entirely of parenchyma, the only primary wood being the circum-medullary strands, which varied in number from five to ten. These, again, were mesarch and branched at regular intervals, one branch turning outwards to become a leaf-trace while the other continued to run vertically up the stem as a reparatory strand.

Eristophyton (Fig. 4C), which may well have been gymnospermous, and *Mesoxylon* (Fig. 4E) were both pycnoxylic (indicated in the diagram by the convention of regular spacing of radial lines) yet the same trend had occurred here, too. In *Eristophyton* the primary wood had become reduced

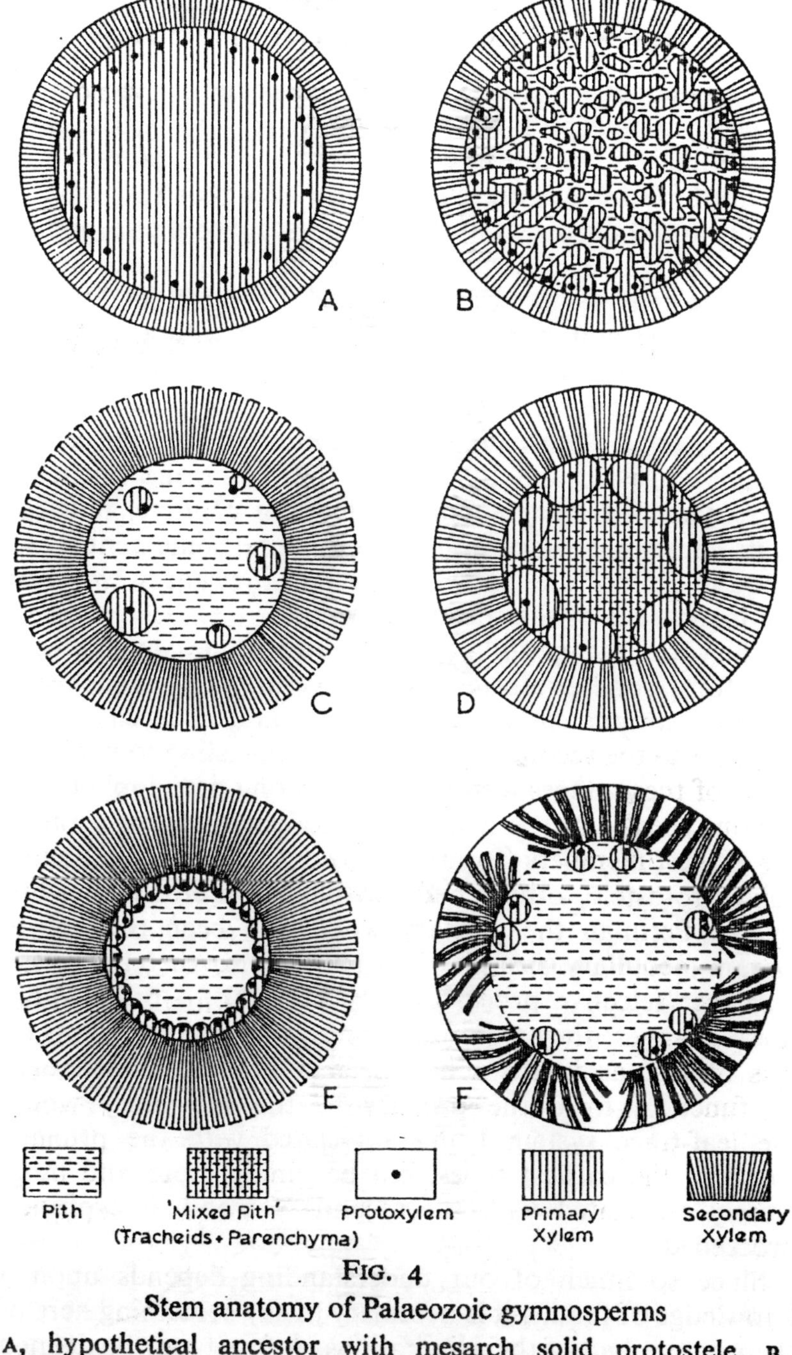

FIG. 4

Stem anatomy of Palaeozoic gymnosperms

A, hypothetical ancestor with mesarch solid protostele. B, *Heterangium Grievii*. C, *Eristophyton*. D, *Calamopitys americana*. E, *Mesoxylon*. F, *Lyginopteris oldhamia*.

to a few circum-medullary strands surrounding the central parenchymatous pith. Most of those visible in any particular cross-section were not in contact with the secondary wood, but, just before turning outwards to become a leaf-trace, a strand would move out until it lay adjacent to the secondary wood. At this level it was mesarch, but when traced downwards it may be seen to have lost its centripetal wood, gradually becoming endarch before fusing with an adjacent strand. Different stages in this process are represented in the single transverse section (Fig. 4C). A related genus, *Sphenoxylon*, showed a more primitive condition, in that the central regions consisted of a mixture of parenchyma and tracheids. *Mesoxylon* was very similar to modern conifers, differing in its somewhat wider pith and in having some centripetal wood in each of the many tiny strands of primary wood on the inner sides of the secondary wood.

The various stages in reduction of the primary wood were, therefore: solid protostele, mixed xylem and parenchyma, mesarch strands surrounding a pure pith, endarch strands adjacent to the secondary wood. It is interesting to note that some of these stages also occurred among pteridophytes, in various species of the woody Lepidodendrales, for some of the earliest members (from the Lower Carboniferous) had a solid protostele, others had medullated protosteles, while by the Permian the primary wood had been reduced to circum-medullary strands. It would seem, therefore, that whatever selective processes were at work to bring about these changes, they were operating in all woody plants. It was almost as if there had been a gradual transference of function from the primary to the secondary wood, the leaf-trace system being associated with the primary wood in the earliest types, but becoming more and more closely associated with the secondary wood as evolution proceeded.

Since so much of our understanding depends upon a knowledge of fossils, it is necessary to say something here of their nature and of their limitations. A fossil may be defined as anything which gives evidence that an organism once existed at some time in the past. Such a wide definition is necessary in order to include 'impressions' (fossils in which no parts of the organism remain, but where just the outline of its shape may still be discerned). Obviously, fossils of this kind are of much less value to the palaeobotanist than are those in which some part of the plant is preserved, however poorly. The best fossils for this purpose are those known as petrifactions, where the plant substance became infiltrated by some chemical, such as calcium carbonate or silica. In the best examples of this kind of preservation the anatomical structure of the original plant can often be studied in fine detail. In the early days of palaeobotany the technique consisted of cutting thin slices of the rock and then grinding them until they were thin enough to be transparent. More recently, it has been found that, on etching the polished surface of the rock with the appropriate acid, the cell-walls remain standing above the surface; these may then be embedded in a film of cellulose acetate, which may be stripped from the rock and examined as if it were a hand-cut section, the thickness depending on the time of etching.

Unfortunately, perfect preservation is rare, for usually some degree of rotting is found to have occurred before impregnation was complete. The soft tissues, such as phloem, were the first to decay, while the lignified and suberised elements were the most resistant. Much more, therefore, is known about the xylem and the bark of petrified plants than of other regions.

In the absence of petrifying minerals, but in anaerobic conditions, plant materials may change gradually into coal. If fossils of this kind are treated with perchloric acid it is possible to oxidise the coally substance away, leaving the cuticles of spores and leaves to be studied under the microscope. This process is easier to carry out in specimens where the various pieces of plant material are separated from each other in the rock by a matrix of sedimentary material. Such fossils are sometimes known as 'mummifications', but more commonly nowadays as 'compressions', and rocks containing them may often be split by a light blow from a hammer, so that the complete specimen is exposed on one half, while an exact counterpart appears as an impression on the other half. Leaves are especially common in the form of compressions, and often provide excellent cuticle preparations. However, internal structures cannot be studied in compressions, except with great difficulty, nor is it always easy to deduce the correct shape of the original plant organ, before it became distorted by pressure. Halle[86] and Thomas[175] are notable for having overcome these difficulties to some degree, for they were able to soften the tissues sufficiently to embed them for sectioning on a microtome.

Perhaps the most difficult problems which confront the palaeobotanist arise from the fact that his specimens, even if their preservation is perfect, usually represent no more than mere fragments of the whole plant. How is he, then, to know which fragments belong to which, for he must know this before he can begin to reconstruct the whole plant in his imagination? The unfortunate truth is that, so far, very few fossil plants indeed have been so reconstructed. In the meantime, all the various bits and pieces must be described under separate generic names ('form-genera'). Thus, the following are now believed to belong to one and the same plant: *Lyginopteris oldhamia* (stems), *Sphenopteris Hoeninghausii* (leaves), *Kaloxylon Hookeri* (roots), *Lagenostoma Lomaxi* (seeds), and *Crossotheca Hoeninghausii* (pollen-bearing organs). The reasons for believing this are, firstly, the frequent association of these structures in the rocks (i.e. where one is present, the others tend also to occur) and, secondly, the fact that some of them are covered with the same peculiar kind of glandular hair. However, the only really convincing proof of identity is the discovery of organic connection between the various parts, and in this instance it is still lacking in the case of the reproductive organs. Thus, although attempts have been made to draw reconstructions of the whole plant, they are still to some extent imaginary, and for this reason some palaeobotanists are not yet prepared to give a scientific name to the whole plant. Others, however, believe that our knowledge is sufficient to justify the name *Calymmatotheca Hoeninghausii* being applied to it (*Calymmatotheca* being the form-generic name for fronds bearing cupulate reproductive organs).

It will be remembered that our definition of the gymnosperms was based entirely on the possession of a seed. Supposing, therefore, that the leaves, roots and pollen-bearing organs associated with *Lyginopteris* stems had not been found in association with seeds, there would be no certainty that they should be classed among the gymnosperms. Yet this is precisely the state of affairs which exists regarding the vast majority of fossil remains. Some may show similarities with other fossils that are known to be gymnosperms; others may show similarities with fossils known to be pteridophytes; but, until reproductive organs are discovered in organic connection with them, their taxonomic placing remains entirely arbitrary.

From what has been said, it will already be clear to the reader that fossil plants present problems to the taxonomist which are quite different from those presented by living plants. The classification of the latter is based on the totality of characters that, together, constitute the whole organism. But, for fossil plants, it would seem that the most convenient procedure would be to have a separate classification for stems, another for leaves, another for seeds, and so on. Such 'organ classifications' do, indeed, exist, and they are a necessary basis for any system of identification. Ultimately, however, the morphologist hopes to see fossil plants classified along with living ones, in such a way as to illuminate their evolutionary history.

The scheme of classification on which this book is based is substantially the same as that proposed by Pilger and Melchior in the 1954 edition of Engler's *Syllabus der Pflanzenfamilien*:[15]

GYMNOSPERMS

A. CYCADOPSIDA
1. **Pteridospermales***
2. **Bennettitales***
3. **Pentoxylales***
4. **Cycadales**

B. CONIFEROPSIDA
1. **Cordaitales***
2. **Coniferales**
3. **Taxales**
4. **Ginkgoales**

C. GNETOPSIDA
1. **Gnetales**

Except for the Gnetales (a remarkable group, which has practically no fossil record and which is rather far removed from the rest), these nine orders of gymnosperms fall into two main groups, on the basis of wood-anatomy, leaf-form and seed-structure. Those belonging to the Cycadopsida tend to have manoxylic wood, large frond-like leaves (basically pinnate in form or in venation) and seeds with radial symmetry. Those belonging to the Coniferopsida have pycnoxylic wood, needle-shaped, paddle-shaped or fan-shaped leaves (basically dichotomous in form or in venation), and seeds which have a bilateral symmetry. This correlation between wood, leaf and seed is not without some exceptions, but it is strong enough to have led some morphologists to conclude that the gymnosperms are diphyletic (i.e. that the two main groups evolved from different pteridophyte ancestors). Since the only feature distinguishing gymnosperms from pteridophytes is the possession of seeds, a belief in a diphyletic origin of the gymnosperms is tantamount to a belief that the seed evolved separately in the two groups.

This view is somewhat strengthened by the fact that both the Cycadopsida and the Coniferopsida appear almost simultaneously in the fossil record. Some idea of the history of the gymnosperms may be gained from Fig. 5, which indicates in a rough way the relative abundance of the various orders during successive geological periods. It will be seen that both the Pteridospermales and the Cordaitales

* An asterisk is used throughout to indicate fossil groups.

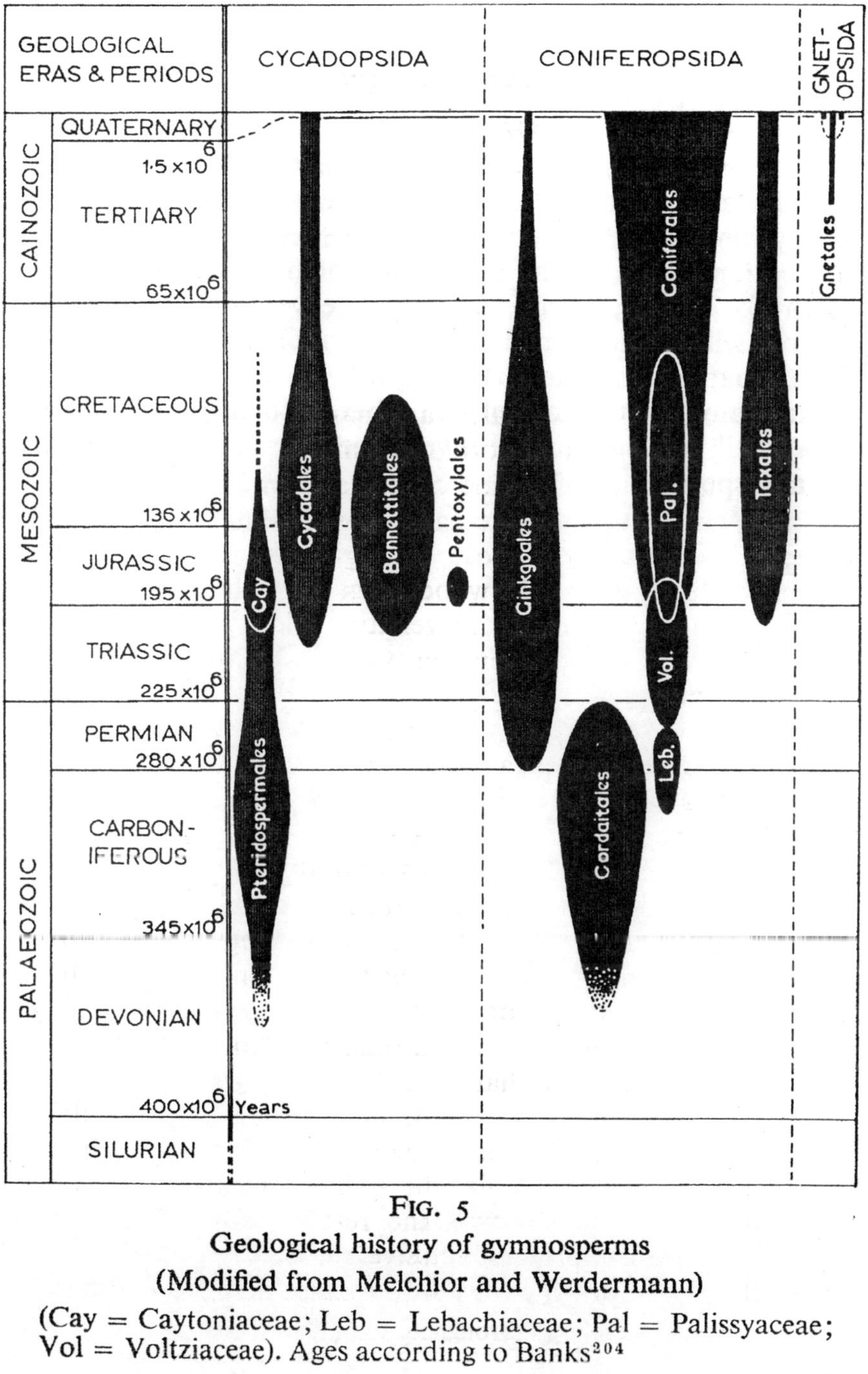

FIG. 5

Geological history of gymnosperms

(Modified from Melchior and Werdermann)

(Cay = Caytoniaceae; Leb = Lebachiaceae; Pal = Palissyaceae; Vol = Voltziaceae). Ages according to Banks[204]

first appeared in the Lower Carboniferous (or possibly in the Upper Devonian), the implication being that neither is the ancestor of the other. In these circumstances, exceptions to the correlation between wood, leaf and seed assume particular importance, for they provide evidence for the contrary view, that the gymnosperms are monophyletic. Several exceptions, such as the association of flattened seeds with fern-like fronds, or of needle-like leaves with manoxylic wood, will be mentioned in subsequent chapters and, in the final chapter, their implications will be discussed in more detail, along with those of a number of fossil-genera which may well represent links between the pteridophytes and the gymnosperms.

2

Palaeozoic Pteridospermales

The Pteridospermales were a very large and diverse assemblage of plants which first appeared in Upper Devonian times and extended through the Carboniferous and Permian to the Mesozoic. It is difficult to define them precisely, not only because of their great diversity, but also because of the fragmentary nature of their fossil remains. Very few are completely known. Descriptions must, accordingly, be based on separate parts which, for the time being, are placed in a large number of form-genera. The following very generalised definition must, therefore, suffice:

> Plants with relatively slender stems. Primary xylem mesarch (rarely exarch) in the form of a solid or a medullated protostele or reduced to circum-medullary strands. Sometimes polystelic. Secondary wood limited in amount, manoxylic and composed of tracheids with multiseriate pitting, especially on the radial walls. Leaves mostly large and fern-like, often many times pinnate. Ovule and seed borne either on the frond or on a specially modified frond ('megasporophyll') which is not part of a cone.

The group may be subdivided into seven families as follows:

1. **Lyginopteridaceae***
2. **Medullosaceae***
3. **Calamopityaceae***
4. **Glossopteridaceae***

5. **Peltaspermaceae***
6. **Corystospermaceae***
7. **Caytoniaceae***

Of these, the first three are confined to rocks of Palaeozoic age and will be described in this chapter. The last four are either confined to the Mesozoic or extend from the late Palaeozoic into the Mesozoic and will be described in a later chapter.

Palaeozoic fern-like fronds

Before dealing in detail with the three families of Palaeozoic pteridosperms it is necessary first to consider the many genera and species of fern-like fronds that, until recently, led botanists to describe the Carboniferous as 'the age of ferns'. Some of these we know to have belonged to ferns because they have been found with fern-like sporangia attached to the margin or to the abaxial side of the pinnules. Others, however, had seeds attached to them, and were equally certainly the fronds of gymnosperms. Some had pollen-bearing organs attached to them, while others have been found in organic connection with woody stems. These, too, are assumed to have belonged to gymnosperms, although it must be emphasised that it is not always easy to distinguish the sporangium of a homosporous fern from the pollen-bearing microsporangium of a gymnosperm. The vast majority of Carboniferous fronds, however, are without reproductive organs and their affinities are, therefore, open to doubt; but it is generally believed that most of them belonged to gymnosperms.

Many of these fronds were so large that their fossils are rarely found intact. Usually they have to be identified from small portions of the frond, in which minute details of shape and venation of the pinnules are the only available characters. A large number of form-genera have, accordingly, been established on the basis of pinnule morphology but, unfortunately, they are not always as sharply defined as one might wish, and they tend to merge with one another.[3] Thus, *Pecopteris* is defined as having had pinnules with

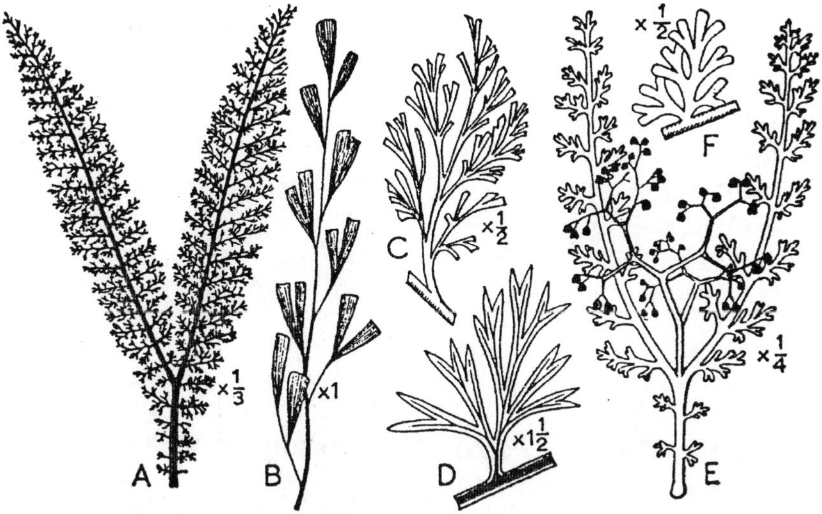

FIG. 6

Lower Carboniferous pteridosperm fronds

A, *Sphenopteridium capillare*, complete frond. B, *Adiantites machanekii*, fragment of frond showing form of ultimate segments. C, *Sphenopteris affinis*, portion of frond. D, *Rhodea Smithii*, ultimate pinna. E, *Diplopteridium teilianum*, reconstruction of complete frond with *Telangium* type of pollen-bearing organs attached. F, pinna of *Sphenopteris* type, from *Diplopteridium teilianum*.

(A, B, E, F after Walton; C, Seward; D, Kidston)

parallel or slightly curved margins attached by the whole width at the base, with a single vein extending as a kind of midrib almost to the tip. The pinnules were free from each other as in *Pecopteris Daubrei* (Fig. 7D) or partly joined laterally as in *P. Armasi* (Fig. 7E) with the result that the pinnae were pinnatifid, rather than pinnate. It is a short step from this type of pinna to that of *Alethopteris* (Fig. 7I) which was also pinnatifid, but differed in that the concurrent portion received veins from the rachis as well as from the pinnule midrib. The form-genus *Callipteridium* merged with *Pecopteris* and *Alethopteris* on one hand, and with *Odontopteris* (Fig. 7G) on the other. (*Odontopteris* had decurrent

FIG. 7

Upper Carboniferous pteridosperm fronds

A, *Mariopteris*, pinna showing double dichotomy. B, *Mariopteris* pinnule. C, *Sphenopteris* pinnule. D, *Pecopteris* (*Asterotheca*) *Daubrei*, pinnule. E, *Pecopteris Armasii*, pinnule. F, *Odontopteris*, complete frond. G, *Odontopteris* pinnule. H, *Alethopteris*, apex of frond. I, *Alethopteris* pinnule. J, *Lonchopteris* pinnule. K, *Neuropteris*, apex of frond. L, *Neuropteris* pinnule. M, *Linopteris* pinnule. (A, B after Zimmermann; C, G, Arnold; D, E, H, Kidston; F, M, Zeiller; I, J, L, Bertrand; K, Stopes)

pinnules with arched veins entering directly from the rachis, but in some forms these veins were more or less aggregated into a midrib.)

The situation is further complicated by the fact that in one and the same frond the pinnule morphology may have varied from place to place. This is well illustrated in Figs. 7H and K, which show the apical regions of an *Alethopteris* frond and of a *Neuropteris* frond, respectively. The fronds of modern ferns show the same kind of variation, where the pinnae become progressively less compound towards the apex, often changing from pinnate through pinnatifid to entire. Fig. 7F, showing the form of the complete frond of *Odontopteris*, illustrates yet another kind of variation, where the pinnules attached to the lowermost regions of the rachis were completely different in shape from those higher up. In a similar way, *Neuropteris* fronds may have had large pinnules, known as *Cyclopteris*, attached to the lower regions of the rachis.

Pecopteris illustrates, better than most, the artificial nature of these form-genera, for some species, e.g. *P. Daubrei* (Fig. 7D), are sometimes found with *Marattiaceous sporangia* on the abaxial surface, and are then placed in the genus *Asterotheca.* Other species, when fertile, bore sori known as *Scolecopteris* and *Ptychocarpus*, while *Pecopteris Pluckenetii* had seeds attached to its pinnules.

When the complete frond is found it is frequently seen to have forked near the base into two equal halves, as shown in Fig. 6A (*Sphenopteridium*), Fig. 6E (*Diplopteridium*) and Fig. 7F (*Odontopteris*). A similar basal dichotomy is recorded in *Neuropteris, Alethopteris, Calymmatotheca, Sphenopteris, Aneimites, Adiantites, Diplothmema, Palmatopteris, Callipteris* and *Eremopteris*.[192] It would seem, therefore, to have been characteristic of pteridosperm fronds, being absent from fronds, such as *Rhacopteris*,[192] which probably belonged to true ferns. In *Diplopteridium* (Fig. 6E) the dichotomy was a 'false' one, because in its angle there was a system of fertile branches, and Walton has suggested that this may well have been a common arrangement in other pteridosperms, too. *Mariopteris* (Fig. 7A) was peculiar in that each pinna underwent a double dichotomy.

No attempt will be made here to list the definitive

characters of each form-genus of fronds. Instead, the reader is referred to the standard textbooks of palaeobotany.[1,3,19,20,23] To the student of gymnosperm evolution it is more instructive to compare the general features of Lower Carboniferous fronds with those of the Upper Carboniferous and the Permian, for they provide strong support for Zimmermann's 'Telome Theory',[25,202] in so far as it attempts to explain the evolution of megaphylls. According to this theory, all lateral organs of vascular plants, both sterile and fertile, have evolved from dichotomous branch systems, and five 'elementary processes' suffice to account for all the changes that have taken place. These are represented diagrammatically in Fig. 8. The starting point is a plant like *Rhynia* or *Horneophyton* (early pteridophytes of Devonian age) in which the branches dichotomised in planes successively at right angles (Fig. 8A). 'Planation' (1) involved a change from three dimensions to two. By a process of 'Overtopping' (2) the dichotomies became unequal and eventually gave rise to a pinnate arrangement, with main and lateral branches. 'Syngenesis', or 'Webbing' (3), increased the photosynthetic area and led to a flattened lamina with dichotomous vena-

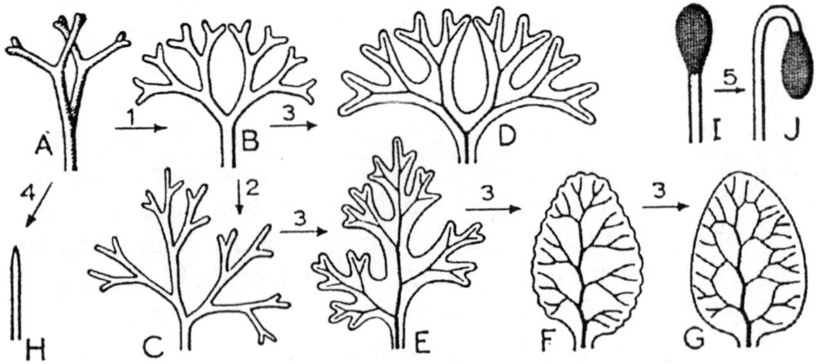

FIG. 8

The Telome Theory

(After Zimmermann, with some modifications)

1, planation; 2, overtopping; 3, syngenesis; 4, reduction; 5, re-curving

tion (Fig. 8D) if this process preceded overtopping, and to one with pinnate venation (Figs. 8E and F) if it came after overtopping. A further stage in syngenesis then resulted in the interconnection of veins to give a reticulate venation (Fig. 8G). These are the three elementary processes that are necessary to explain the evolution of megaphylls. Microphylls, on the other hand, are held to be the result of a fourth process, 'Reduction' (4). Lastly, in the evolution of many reproductive organs, 'Re-curving' (5) has occurred.

The fronds and pinnules illustrated in Fig. 6 are all from the Lower Carboniferous. The most noticeable feature is the large number of species with much-branched pinnules in which the ultimate segments were still narrow and axis-like,[177] e.g. *Sphenopteridium* (Fig. 6A). *Rhodea* (Fig. 6D) had scarcely begun to exhibit syngenesis, while the fertile portion of *Diplopteridium* was at an even earlier stage and branched in three dimensions. Other Lower Carboniferous pinnules were slightly expanded, being spathulate or deltoid, e.g. *Adiantites* (Fig. 6B), but the venation was dichotomous. Hardly any had compact fronds with broad, closely-set pinnules, such as are illustrated in Fig. 7, from the Upper Carboniferous. One of the few genera to occur in both the Lower and the Upper Carboniferous was *Sphenopteris*, but even in this genus there were marked differences (cf. Figs. 6C and 7C). In many Upper Carboniferous pinnules there was a well-marked midrib with lateral veins, while several genera had reticulate venation, e.g. *Lonchopteris* (Fig. 7J) and *Linopteris* (Fig. 7M). By Lower Permian times there were large fronds so highly advanced in their form and venation as to be comparable to the leaves of dicotyledons (*Gigantopteris* was, indeed, compared to *Nicotiana* some years ago).

Lyginopteridaceae*

Stems: *Tetrastichia, Tristichia, Rhetinangium, Heterangium, Lyginopteris, Callistophyton, Schopfiastrum*

Fronds: *Sphenopteris*, etc

Seeds (Lower Carboniferous): *Sphaerostoma, Salpingostoma, Calathospermum, Geminitheca, Genomosperma, Eccroustosperma, Hydrasperma, Stamnostoma, Eosperma, Dolychosperma, Lyrasperma, Eurystoma, Deltasperma, Camptosperma* (Upper Carboniferous): *Lagenostoma, Physostoma, Conostoma, Tyliosperma*

Pollen-bearing organs: *Telangium, Crossotheca,* etc

This group was represented in both the Upper and the Lower Carboniferous. The plants belonging to it had monostelic stems that cannot have been strong enough to support the weight of their large fronds. If not actually climbers, therefore, they must have had a straggling growth habit. Their seeds, with few exceptions, had the integument fused with the nucellus except at the apex, and many are known to have been borne in a cupule.

There are two monotypic genera, *Tetrastichia* and *Tristichia*, from the Lower Carboniferous of Scotland, whose primary stem structures would almost certainly have placed them with the ferns were it not for their secondary wood and the cortical bands of sclerenchyma, which are typical of pteridosperm stems. A transverse section through the stele of *Tetrasticha bupatides*[77] is illustrated, somewhat diagrammatically, in Fig. 9A. The stem was less than 1 cm in diameter and branched in an opposite and decussate fashion. The primary xylem was cruciate (or occasionally five-armed) and consisted of a solid protostele of scalariform or reticulately pitted tracheids, without any admixture of parenchyma. About midway along each arm was a mesarch protoxylem group from which the branch-trace protoxylems had their origin. Gordon regarded these lateral branches as petioles; there was a swollen pulvinus at the base of each and the vascular bundle entering it was butterfly-shaped. The 'petiole' remained unbranched for about 12 cm and then bifurcated, like most pteridosperm fronds, but what kind of frond was borne beyond this point is not known with certainty (some believe it to have been *Sphenopteris affinis*).

Some branches were without secondary thickening, but others had a narrow zone of secondary wood that followed the outline of the primary wood. This was made up of tracheids with reticulate thickenings and both uniseriate and multiseriate wood-rays. In the inner cortex there were scattered groups of stone-cells ('sclerotic nests') and in the outer cortex a network of plates of fibres ('Sparganum cortex').

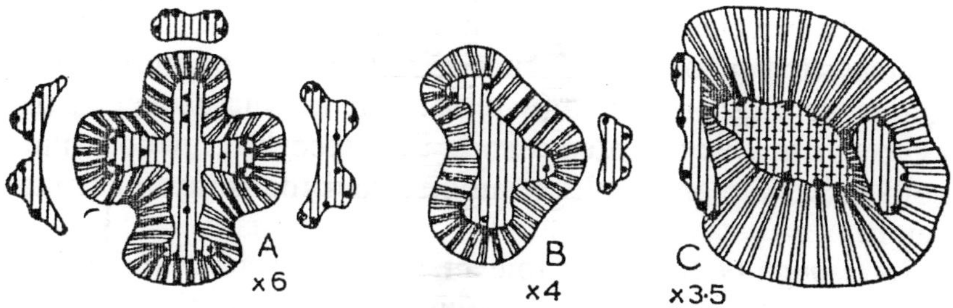

FIG. 9

Vascular systems of primitive pteridosperms

A, *Tetrastichia*. B, *Tristichia*. C, *Schopfiastrum*.

(A, based on Gordon; B, on Long; C, on Andrews)

(For key see Fig. 4)

Tristichia Ovensii (Fig. 9B)[123] was similar in many respects to *Tetrastichia*, differing mainly in its ⅓ phyllotaxy. In keeping with this difference, its primary wood had three radiating arms instead of four. The 'petioles' received a butterfly-shaped trace, they dichotomised near the base and then dichotomised again repeatedly. The secondary wood had multiseriate pits, mostly on the radial walls, and the cortex had sclerotic nests and a 'Sparganum' hypoderma. Associated with this fossil were seed-bearing cupules belonging to the genus *Stamnostoma* (Fig. 10H). A suggestion recently put forward by Long[124] is that these axes represent the fertile continuation of the rachis of a frond, the 'petioles' being no more than lateral pinnae, and it may well prove ultimately that *Tetrastichia* was similar in this respect.

Rhetinangium Aberi,[75] also from the Lower Carboniferous of Scotland, is known from three pieces of petrified stem and several petioles. The stem was about 2 cm in diameter, with a central stele about 7 mm across. The primary xylem was exarch and consisted of numerous strands of tracheids separated by a meshwork of parenchyma. The inner cortex was without sclerotic nests, but in the outer cortex there were vertical plates of fibres that interconnected at distant intervals. The vascular supply to the petiole was unlike that of any other pteridosperm in that several of the peripheral strands of xylem turned outwards, becoming U-shaped in the process, and remaining loosely attached, so as to form a corrugated band. The secondary wood had both uniseriate and multiseriate heterogeneous rays of great height.

Heterangium Grievii is one of the commonest and most easily recognisable fossil stems in the calciferous sandstone of Pettycur, Fifeshire. Branches up to 5 cm in diameter are known, but even in the largest the stele was only about 15 mm across and had only a very narrow zone of secondary wood. Its structure is indicated, very diagrammatically, in Fig. 4B. The primary xylem was mesarch and protostelic, but the tracheids were separated into strands by parenchyma as in *Rhetinangium*. The secondary wood had tracheids with multiseriate pits on the radial walls, and wood-rays like those of *Rhetinangium*. There were horizontal plates of stone-cells in the inner cortex and parallel vertical bands of fibres in the outer cortex. In the formation of a leaf-trace one of the peripheral mesarch strands turned outwards, at first with just a single protoxylem but later with two. The petioles were attached in a $\frac{2}{5}$ or $\frac{3}{8}$ phyllotaxy. What the fronds were like is not known with certainty, but the stems are frequently associated with fronds called *Diplothmema adiantoides*, which forked once near the base and bore *Sphenopteris* pinnules. The fact that the stems are also often found in association with small seeds, known as *Sphaerostoma ovale* (Fig. 10A), has led to the belief that they belonged to one and the same plant, but organic connection between the two has yet to be demonstrated. A detailed description of *Sphaerostoma* will be deferred until later when, along with other genera, the seeds of the Lower Carboniferous will be compared with those of the Upper Carboniferous.

Some twelve species of *Heterangium* are known from the Upper Carboniferous, placed by Scott[19] in the three subgenera *Eu-Heterangium*, *Polyangium* and *Lyginangium*, which are believed to represent successive levels of evolution. The first contains species which are like *H. Grievii* in that the leaf-trace was single at its point of origin. In *H. Schusteri* the leaf-trace divided into two almost immediately on leaving the stele. From this species it is a short step to those of the *Polyangium* group, in which the leaf-trace was a double structure from the start and underwent further subdivisions on its way out to the leaf-base, e.g. *H. tiliaeoides* and *H. shorense*.

The *Lyginangium* group contains species in which some degree of medullation had taken place. Thus, although in *H. Andrei* the xylem tracheids still extended to the centre, parenchyma predominated in this region. In this respect it showed a close approach to *Lyginopteris*, for in *L. heterangioides* there was a central pith region with a few scattered tracheids, as a vestige of the ancestral protostelic condition. In all other species of *Lyginopteris* the process of medullation was complete and the primary wood was represented merely by the peripheral leaf-trace system.

Lyginopteris oldhamia is extremely common in coal-balls from the coal-mines of Lancashire and Yorkshire, and is known in greater detail than any other fossil plant. The stem[19] attained a diameter of 4 cm and branched frequently. Its vascular system, illustrated in Fig. 4F, consisted of a number (5–10) of peripheral mesarch primary xylem strands surrounding a central pith region, and in all but the narrowest axes there was a zone of secondary wood. This was of a very loose kind, because the multiseriate rays, although fairly narrow in the first-formed secondary wood, were greatly dilated towards the outside. This feature has been indicated diagrammatically in Fig. 4F. Uniseriate rays were extremely rare. The pith region contained numerous 'sclerotic nests', as did the pericycle also. The outer regions of the pericycle tended to become meristematic, giving rise to a phellogen, by whose activity a zone of periderm was produced on its outer side. In the outer cortex there was a very well-marked system of longitudinal plates of fibres which anastomosed to form a regular network ('dictyoxylon cortex'). In transverse section these give the general impression of the Roman numerals on a clock face and make *Lyginopteris* one of the easiest fossils to recognise.

The leaves were borne in a $\frac{2}{5}$ phyllotaxy, and in the formation of a leaf-trace one of the peripheral xylem strands split into two. Of these, one continued up the stem as a reparatory strand, while the other turned outwards towards the leaf-base. At first the leaf-trace had a single protoxylem, but this divided into two before the strand started to turn outwards. On its way out, the trace split into two, but these fused into single strand again on entering the base of the petiole, the number of protoxylems increasing meanwhile. Within the petiole, the vascular strand became V-shaped and was not unlike the butterfly-shaped trace of *Tetrastichia*. The frond, known as *Sphenopteris Hoeninghausii*, forked once at a point just above the basal pinnae and, in preparation for this, the V-shaped trace became W-shaped before separating into two.

That these fronds really belonged to *Lyginopteris* is confirmed, not only by the continuity of the dictyoxylon cortex into the petiole, but also because of the occurrence on both stem and petiole of peculiar glandular 'spines'. Indeed, these characteristic spines occurred on all parts of the plant except the roots and establish beyond all reasonable doubt that its seed was the well-known *Lagenostoma Lomaxii* (Figs 10 O and P), for the cupule surrounding the seed was also covered with such glandular spines. Concerning the pollen-bearing organs of *Lyginopteris*, however, there is rather less certainty, although it is generally accepted that they were probably of the type known as *Crossotheca* (to be described later).

Although *Lyginopteris* is not represented in North American deposits, *Callistophyton poroxyloides*, first described in 1954,[54] was similar to it in many respects. Its stems attained a diameter of just over 2 cm and are so well preserved that even the sieve-plates on the lateral walls of the phloem sieve-cells are visible. In the centre was a parenchymatous pith, around which were several mesarch peripheral strands of primary wood. Axillary branching was a well-marked and interesting feature, the fronds (and branches) being borne in a spiral sequence. The secondary wood was somewhat more compact than that of *Lyginopteris*, because the wood-rays were narrower (up to seven cells wide) and the bands of fibres in the outer cortex did not anastomose. Little is known, unfortunately, of the apical regions of most fossil plants because of the extreme delicacy of the cell-walls in meristematic regions. It was, therefore, of the greatest interest that a beautifully preserved stem apex of *Callistophyton* was described in 1956,[51] and the young leaves were shown to have been coiled like those of a fern ('circinate vernation').

Schopfiastrum decussatum, also from U.S. coal-balls,[28] was at a lower stage of evolution, in that its primary stem structure was little more than an exarch rod of tracheids. There were, it is true, some parenchyma cells, but far fewer than in the Lower Carboniferous *Rhetinangium* or *Heterangium*. Fig. 9C shows the way in which leaf-traces were given off in an opposite and (presumably) decussate fashion, and illustrates also their comparatively large size. In the outer cortex there were the usual fibrous strands, which anastomosed only at extended intervals, appearing to be nearly parallel.

Seeds of the Lyginopteridaceae have been known for a long time, and new genera and species are being discovered and described each year. Recent discoveries in Scotland have considerably widened our knowledge of Lower Carboniferous seeds, and have thrown a new light on the nature and evolution of their various parts.[122]

Sphaerostoma ovale (Fig. 10A), believed to be the seed of *Heterangium Grievii*, was originally described in 1877 under the name *Conostoma ovale*, but was re-examined by Benson in 1914.[43] Like all the seeds of the Lyginopteridaceae, it was relatively small (3·5 mm × 2 mm). During its early development it was closely invested by a cupule (3), but at maturity it was shed without its envelope, and it was in this condition that most of them became fossilised. The one illustrated in Fig. 10A was still within its cupule. This had a number of vascular bundles (probably eight) running vertically within it. The seed also had eight vascular bundles within its integument, which was fused to the nucellus except right at the apex, where it formed a 'canopy' over the nucellar cap. Round the micropyle were eight small crested lobes of the canopy corresponding with the eight vascular bundles.

The nucellar cap was modified into a rather complicated pollen-capturing device—the 'lagenostome'. Figs. 10B–D illustrate the way in which it may have operated. It is thought that, to begin with, the nucellar cap was a solid structure comprising a flat plinth (17) and a central dome. Then, by means of a circumscissile split (10), the epidermis separated from the sides of the dome and stretched, so as to allow the pollen-grains to enter (Fig. 10C). These then became trapped (12) by the subsequent shrinkage of this epidermis (now the wall of the pollen-chamber). By then, the floor of the pollen-chamber and the megaspore-wall (7) would have dissolved away so that the archegonia beneath would have been readily accessible. Of the hundreds of seeds examined, none has been found with an embryo inside, so it is assumed that fertilisation must have taken place after the pollinated seed had been shed from the parent plant. This, indeed, is probably what happened in all palaeozoic seeds.

In *Sphaerostoma* each cupule contained just the one seed, and presumably this was true of many other Lower Carboniferous pteridosperms. However, several species are known in which each cupule contained several seeds. Thus *Calathospermum scoticum* (Fig. 10E) was a cupule, shaped rather like a tulip flower 4·5 cm long, which contained a number of stalked seeds. Some have been found with as many as seventy stalks, from which the seeds had become detached, and Walton[193] has suggested that, as each seed reached maturity, its stalk elongated, carrying it up to the mouth of the cupule. Then after pollination it was shed from its stalk, making room for the next seed as it matured in turn.

The 'body' of the seed was 3 × 2 mm, but the nine tentacle-like lobes of the integument increased the overall length of the seed to about 15 mm. Internally it was very similar indeed to *Salpingostoma dasu*[78] (Fig. 10F). This was a much larger seed, 14 × 6 mm, with six 'tentacles' about 2·5 cm long. Both genera are peculiar in that there was a trumpet-shaped extension beyond the lagenostome, called the 'salpinx' (13). Hairs on the inside of the tentacles closely invested this salpinx and, presumably, helped to guide the pollen-grains into it and so down to the pollen-chamber. The tentacles of both genera were free only to the 'shoulders' of the seed; below this level the integument was continuous and was fused to the nucellus. *Hydrasperma tenuis*[122] (Fig. 10I) was similar in this respect. It was about 3·5 × 1 mm, with eight to ten free lobes of the integument diverging like the tentacles of a semi-contracted *Hydra*. The apparatus for pollen reception in this seed was intermediate in some respects between that of *Sphaerostoma* and that of *Salpingostoma*, for the wall of the pollen-chamber was prolonged into a short funnel-like salpinx, but there was a central plug of tissue extending up into it.

Genomosperma[122] was a remarkable seed in that the lobes of the integument were free from each other and from the nucellus right to the chalazal end of the seed. In *G. Kidstonii* (= *Calymmatotheca Kidstonii*) (Fig. 10G) the overall length of the seed was from 10 to 15 mm, the number of lobes varied from six to eleven (the commonest number being

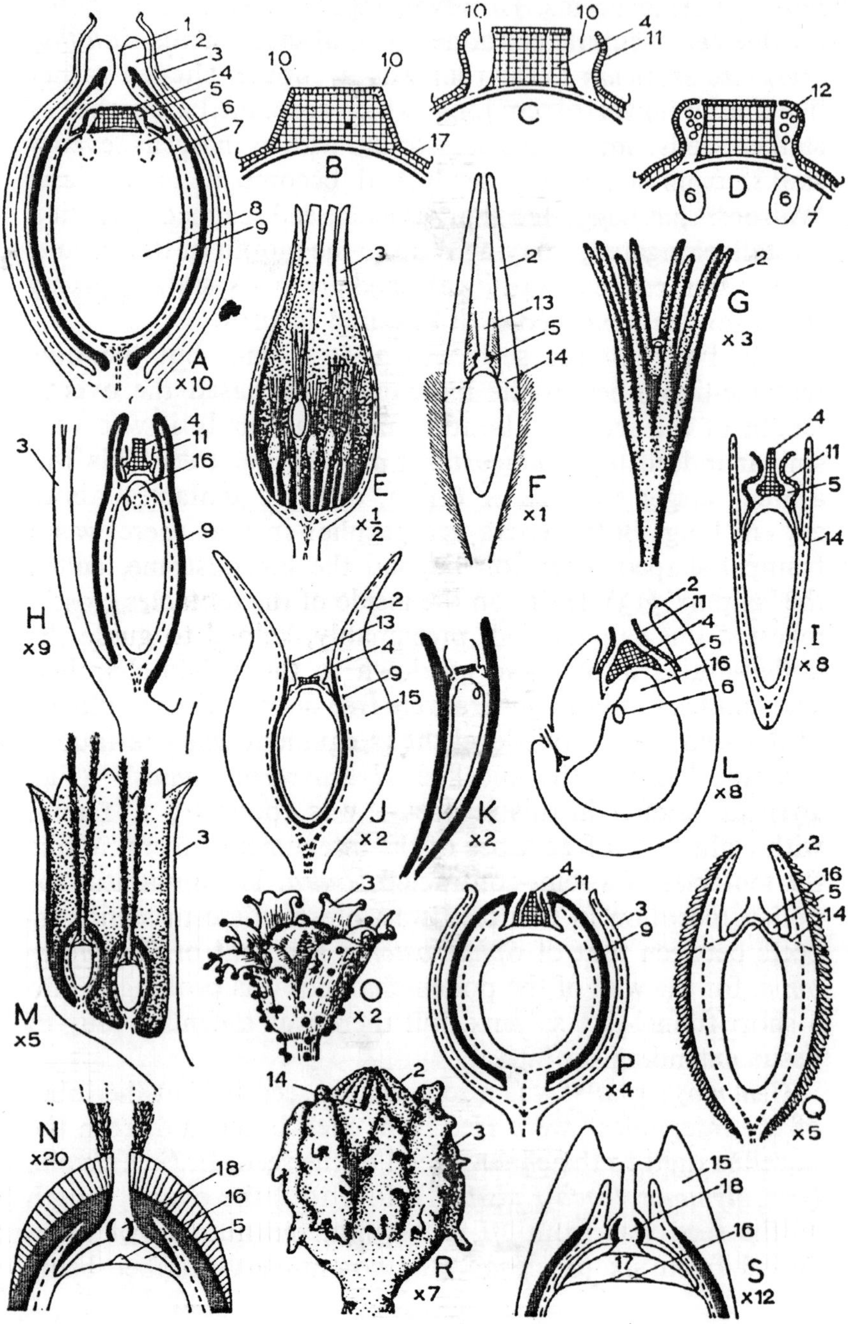

A
x10
1
2
3
4
5
6
7
8
9
B
10
10
17
C
10
10
4
11
D
12
6
6
7
E
x 1/2
3
F
x1
2
13
5
14
G
x 3
2
H
x9
3
4
11
16
9
I
x 8
4
11
5
14
J
x 2
2
13
4
9
15
K
x2
L
x8
2
11
4
5
16
6
M
x5
3
O
x 2
3
P
x4
4
11
3
9
Q
x5
2
16
5
14
N
x20
18
16
5
R
x7
14
2
3
S
x12
15
18
16
17

eight) and the lobes were diverging, whereas in *G. latus* they were loosely converging. There was a minute salpinx at the apex of the lagenstome which had a central plug.

All the seeds described so far were radiospermic, i.e. they had a radial symmetry, but even in Lower Carboniferous times there were some with a bilateral symmetry, i.e. platyspermic. *Lyrasperma scotica*[122] (Figs. 10J and K) is a good example of this kind of symmetry, for not only was it bilateral in external symmetry, but internally also, having had only two vascular bundles in its integument. These ran up in the major plane (Fig. 10J) beneath the wings of the seed, which were continued into diverging horns at the apex.

FIG. 10

Seeds of Lyginopteridaceae

(A–L from Lower Carboniferous; M–S from Upper Carboniferous)

A, *Sphaerostoma ovule*, l.s. seed in cupule (B, C, D, stages in development of the nucellar cap). E, *Calathospermum scoticum*, cupule cut in half to show stalked seeds inside. F, *Salpingostoma dasu*, l.s. G, *Genomosperma Kidstonii*, reconstruction to show free integument-lobes. H, *Stamnostoma huttonense*, l.s. showing part of cupule-lobes. I, *Hydrasperma tenuis*, l.s. J, K, *Lyrasperma scotica*, l.s. seed in planes at right angles. L, *Camptosperma berniciense*, l.s. showing campylotropous shape. M, *Gnetopsis elliptica*, l.s. cupule with two seeds, partly hypothetical (N, l.s. micropylar region). O, *Lagenostoma Lomaxii*, reconstruction of seed in cupule (P, l.s. of seed and cupule). Q, *Physostoma elegans*, l.s. R, *Tyliosperma orbiculatum*, reconstruction of seed, showing fleshy cupule-lobes. S, *Conostoma anglo-germanicum*, l.s. micropylar region.

(1, micropyle; 2, integument-lobes; 3, cupule; 4, central plug; 5, pollen-chamber; 6, archegonium; 7, megaspore-wall; 8, female prothallus; 9, integumentary vascular supply; 10, line of circumscissile dehiscence; 11, wall of pollen-chamber; 12, pollen-grains; 13, salpinx; 14, level at which integument-lobes become free from each other; 15, wing-like flange of integument; 16, papilla, or tent-pole; 17, plinth; 18, barrel-shaped lagenostome)

(G, based on information from Long; A, after Benson; E, Walton; F, Gordon; H, I, J, K, L, Long; M, Renault and Zeiller; N, S, Oliver and Salisbury; O, P, Oliver and Scott; Q, Oliver; R, Mamay)

Fig. 10K illustrates the appearance of this seed as if it were cut in a plane at right angles to that of Fig. 10J. So similar is this seed, in external form, to those of the Cordaitales that it was originally described under the name *Samaropsis scotica*. However, its lagenostome with a central plug shows a much closer affinity with the seeds of the Lyginopteridaceae (grouped together by some palaeobotanists as the Lagenostomales). The large barrel-shaped salpinx of *Lyrasperma* is noteworthy. *Eosperma oxroadense*[38] was also platyspermic, 3·3 mm long × 3·3 mm wide × 0·5 mm thick, and in this species, also, the bilateral symmetry extended to the vascular system, for there were only two vascular bundles in the integument. However, there were no wings to the seed, nor was the integument lobed at the apex.

Another remarkable seed, recently described by Long,[122] was campylotropous, i.e. it was curved so that the micropyle and the chalaza were on the same side of the seed. This was *Camptosperma berniciense* (Fig. 10L). Hitherto, seeds of this shape had been known only in a few families of flowering plants. That they should have been represented among the earliest of all seeds is, indeed, astonishing.

The cupules of *Sphaerostoma* and *Calathospermum*, described above, were both radially symmetrical. It is therefore not immediately clear that they formed part of a dorsiventral frond. However, the vascular bundle entering the base of the cupule was dorsiventral. By contrast, the cupule of *Stamnostoma huttonsense*[122] consisted, basically, of a system of relatively simple cylindrical axes that branched dichotomously and were folded round the four seeds, which were attached near the lowermost dichotomies. Fig. 10H illustrates one of these seeds and part of the cupule system. The seed itself was 3·75 × 1·5 mm in size and was barrel-shaped. The integument was not lobed at all, and the most striking feature of all was the micropyle, which was so wide as to be comparable with the open mouth of a jar. Although this seed was cylindrical and un-ridged over most of its length, the number of vascular bundles entering its base points to an underlying hexamerous or pentamerous symmetry. *S. bifrons* was slightly smaller than *S. huttonense* and was similar in shape to *Sphaerostoma ovule*. However, as Long[122] has pointed out, the presence of a wide salpinx makes it impossible to classify it in the genus *Sphaerostoma*.

Long[122] described yet another seed which appears to have been protected to some degree by the folding round it of a system of branches. This was *Eurystoma angulare*, a quadrangular seed with four integument-lobes and a slightly keeled base. The branches that were slightly curved round it appear to have lacked photosynthetic tissue and might therefore be interpreted in a number of different ways: as branches, as a specially modified part of an otherwise photosynthetic frond, or as representing the early stages of evolution of a fertile frond from a telome system.

The variations shown by Lower Carboniferous seeds in the manner in which they were borne in the cupule are paralleled in the Upper Carboniferous, for in *Gnetopsis* it is known that the cupule contained two seeds, while in *Lagenostoma* each seed was enveloped by its own cupule. Some had relatively wide micropyles, but in the majority the micropyle was narrow, and the salpinx seems to have disappeared. The cupule of *Gnetopsis*[17] was about 6 mm long and 3 mm across; it was bilaterally symmetrical and was bilobed with a toothed margin. The seeds within it were slightly flattened and relatively small, being only 2·5 × 1·2 mm, but they were provided with four very long plumes which projected beyond the rim of the cupule (Fig. 10M). Presumably these functioned in much the same way as the hair-lined tentacles of *Salpingostoma* in directing the pollen-grains down to the lagenostome. This was a barrel-shaped structure (Figs. 10N–18), without any central plug, situated at the apex of the plinth, and there is evidence that, after pollination, a 'tent-pole' (16) extended upwards and blocked it, thereby trapping the pollen-grains beneath the plinth. *Conostoma*[138] had a very similar lagenostome, but it lacked the plumes. Furthermore, the seed was somewhat larger, being about 7 × 2·3 mm. *C. anglo-germanica* (Fig. 10S) was eight-angled and had four wings at the apex, reminiscent of the four fins on a bomb, whereas *C. oblongum* was six-ridged, was slightly platyspermic and lacked wings at the apex.

Lagenostoma Lomaxii,[139] the seed of *Lyginopteris Hoeninghausii,* is usually found without its cupule, but during its early development it was invested by a lobed cupule covered with glandular hairs (Fig. 10O). The seed was about 5·5 × 4·25 mm and, although there were about nine vascular bundles within the integument, there were no free lobes at the apex. There was just a simple hole, through which the lagenostome protruded slightly. This was bottle-shaped, with a central plug (Fig. 10P). Long[121] has described a well-preserved seed in which the prothallus, besides containing clear evidence of archegonia, shows a well-marked tent-pole (not illustrated in Fig. 10P).

It is not known whether *Physostoma* was borne in a cupule during its early development, for it has always been found without one. *P. elegans*[137] (Fig. 10Q) was about 6 × 2·25 mm and usually had ten integument-lobes, although some have been found with only nine and some with as many as twelve. It was very different from the other seeds of the Upper Carboniferous in several respects. The lobes of the integument were more like the diverging tentacles of the Lower Carboniferous *Hydrasperma.* The seed was further characterised by the outer covering of large club-shaped epidermal hairs. The entire lack of sclerenchyma from the seed was also peculiar; so was the very simple pollen-chamber and the very large tent-pole that projected into it, replacing, as it were, the central plug. *P. Kidstonii* differed only in the number of integument-lobes (six). *P. stellatum,* too, had six integument-lobes, but the body of the seed was so deeply grooved as to give it a stellate shape in transverse section.[109]

In 1954 Mamay[129] described a very interesting seed, *Tyliosperma orbiculatum* (Fig. 10R), from an American coal-ball of Pennsylvanian age (i.e. Upper Carboniferous). It was almost spherical, about 3·6 mm in diameter, and its seven lobed integument was fused with the nucellus except at the apex. The particular feature of interest lies not in the seed so much as in its cupule, which was divided to the base into seven, or possibly eight, fleshy lobes. It is difficult to imagine that these were part of a frond wrapped round the seed, for they were finger-like outgrowths from the chalazal end of the seed. It would seem, therefore, that there may well have been two quite different kinds of cupule among the Lyginopteridaceae. Radially symmetrical ones, like those of *Lagenostoma* and *Sphaerostoma*, could have resulted from the phylogenetic 'fusion' of a number of separate lobes, while bilaterally symmetrical cupules containing several seeds, like those of *Stamnostoma*, *Gnetopsis* and *Calathospermum scoticum*, could have resulted from the folding round of the ultimate segments of the frond. *Tyliosperma*, then, represented a primitive example of the first type, which was still surviving in the Upper Carboniferous.

As in the case of the cupule, so also the integument has been regarded by some morphologists as a part of the frond wrapped round the seed. Others, aware that some seeds were attached to branch systems which had not yet achieved the status of leaves, regard the integument as having formed by the fusion of a number of branch-tips, i.e. 'telomic'. Andrews[1] recently put forward a radically different theory to account for the integument and its vascular supply. Based, in part, on suggestions by Walton and by Mamay, this theory supposes that the seed evolved from a megasporangium, like that of the fern *Stauropteris burntislandica*, terminating a stout axis. 'Sinking' of the sporangium into the axis resulted in its being surrounded by the vascular system of the axis and, at the same time, the apex of the sporangium is supposed to have become modified for pollen-reception, lobes growing up to produce a micropyle. Ingenious as this theory is, it seems less likely to represent the truth than does the telomic theory, and Andrews himself has now accepted this view.[29]

The discovery of *Genomosperma* provides the strongest

possible evidence for the telomic interpretation and, indeed, represents such an early stage in seed evolution that, as Delevoryas[8] remarks, it scarcely qualifies to be called a seed, for the megasporangium cannot be described as having been enclosed by the integument. The integument-lobes could, of course, be regarded as outgrowths from the base of the sporangium, taking them at their face value and regarding them as arising *de novo*. However, most morphologists are reluctant to accept the idea of plant organs arising *de novo*, and prefer to search for some pre-existing organ from which they could have evolved.

Even if the telomic interpretation should be accepted, however, there still remains the question as to whether the integument-lobes came from sterile or fertile telomes. As long ago as 1904, Benson suggested that the pteridosperm seed should be interpreted as a synangium, in which all but one central sporangium had become sterile. Halle[87] developed this idea further, referring to the Silurian pteridophyte *Yarravia* (Fig. 11J) as the kind of plant from which this evolutionary process might have started. The fertile branches of this plant terminated in a radial group of sporangia partially fused into a synangium. Smith,[163] in an important review of theories of ovule evolution, has emphasised that no intermediate stages between *Yarravia* and *Genomosperma* have yet been found in the fossil record. However, Benson's theory is particularly attractive, because the pollen-bearing organs attributed to the Lyginopteridaceae and to the Medullosaceae are also made up of radial groups of sporangia terminating an axis (Fig. 11). The seed and the male organ can thus be regarded as completely homologous —as, indeed, might be expected, if seed-plants were descended from some homosporous ancestor.

It is much to be regretted that, so far, no pollen-bearing organs have been found in organic connection with stems or seeds of the Lyginopteridaceae, for not a single species can yet be said to be completely known. However, circumstantial evidence strongly supports the belief that *Telangium*, *Schuetzia*, *Alcicornopteris*, *Diplotheca*, *Diplopteridium* and *Crossotheca* all represent such pollen-bearing organs. *Telangium affine* (Fig. 11D) is associated in rocks of Lower Carboniferous age with *Sphenopteris affinis*, and there is reason to believe that it belonged to *Tetrastichia* 'stems'.[1] It consisted of a group of six to eight elongated bilocular sporangia, united near the base into a synangial disc. Dehiscence occurred by means of a split down the inner side of each sporangium, thus liberating the pollen-grains. *Telangium bifidum* (Figs. 11B and C) differed in that there were up to twenty-five sporangia on each disc. *Diplopteridium teilianum* (Fig. 11A) had very similar synangial discs, borne on a slender rachis with wide-angled forkings, attached in the main angle of the forked frond (Fig. 6E). Most fronds of *Diplopteridium* are found without this fertile branch system, but the fact that some specimens show a stump, or a scar, in the angle suggests that the fertile portion may have ripened and dropped off before the vegetative part of the frond had matured. *Schuetzia Bennieana* (Fig. 11H) had pedicillate campanulate synangia, arranged spirally on a fertile axis. The number of sporangia in each synangium is difficult to determine, so tightly were they pressed together, but it was probably between fifteen and twenty. The pollen-grains were very numerous, spherical and smooth, a tri-radiate scar on each suggesting that they were formed in tetrads. *Alcicornopteris* (Fig. 11 I, likewise, had about fifteen sporangia attached to a peltate disc, but *Diplotheca* was somewhat different in that each synangial disc bore ten sporangia united in pairs at the base.

Of these genera, only *Telangium* survived into the Upper Carboniferous. *T. Scottii* was thought for some time to have belonged to *Lyginopteris oldhamia*, but most palaeobotanists now believe that the pollen-bearing organs of this plant are more likely to have been *Crossotheca*, for this genus, too, was borne on *Sphenopteris* fronds (although a few species have also been found attached to *Pecopteris* fronds, e.g. *C.*

sagittata). In *Crossotheca* a number of boot-shaped bilocular sporangia were arranged radially on the underside of discs (Figs. 11E and F). These terminated the ultimate segments

FIG. 11

Pollen-bearing organs of Carboniferous pteridosperms

(A–I, Lyginopteridaceae; K–Q, Medullosaceae; J, possible ancestor)

A, *Diplopteridium teilianum*. B, C, *Telangium bifidum*. D, *Telangium affine*. E, F, *Crossotheca Hoeninghausii*. G, *Crossotheca* sp. H, *Schuetzia* sp. I, *Alcicornopteris*. J, *Yarravia oblonga*, a member of the Psilophytales, from the Upper Silurian/Lower Devonian. K, *Whittleseya elegans*. L, *Codonotheca caduca*. M, N, *Aulacotheca elongata*. O, *Dolerotheca formosa*. P, *Goldenbergia glomerata*. Q, *Boulaya fertilis*. R, *Potoniea adiantiformis*.

(A, after Walton; B, C, E, F, H, I, Kidston; D, based on Benson; G, after Andrews; J, Lang and Cookson; K, L, M, N, Q, R, Halle; O, based on Schopf)

of the fertile frond (Fig. 11G) and have been variously described as resembling minute hairbrushes, or epaulets, because of their lateral attachment. However, apart from their asymmetrical attachment, they were not unlike *Telangium* in their basic organisation. Indeed, one can only conclude that the pollen-bearing organs of the Lyginopteridaceae were much more uniform than their seeds.

Medullosaceae*

Stems: *Sutcliffia*, *Medullosa*

Petioles: *Myeloxylon*

Fronds: *Neuropteris*, *Alethopteris*, etc

Seeds: *Stephanospermum*, *Aethiotesta*, *Pachytesta* (*Trigonocarpus*), *Hexapterospermum*, *Polypterospermum*, *Ptychotesta*, *Polylophospermum*, *Codonospermum*

Pollen-bearing organs: *Codonotheca*, *Whittleseya*, *Goldenbergia*, *Aulacotheca*, *Boulaya*, *Dolerotheca*, *Potoniea*

This group of pteridosperms, which extended from the Upper Carboniferous to the Permian, includes some of the most remarkable and perplexing plants in the vegetable kingdom. They were more massive in all respects than the Lyginopteridaceae, for not only were their seeds and pollen-bearing organs large, their stems (particularly those from the Permian) were more trunk-like and all were polystelic, some specimens having more than forty steles. Perhaps the most perplexing feature is the way in which each of the many steles was able to undergo secondary thickening without disrupting the intervening tissues. Presumably, the ground tissue must have retained the power of meristematic growth throughout the life of the trunk, thereby allowing for adjustments to be made for the enlarging steles. Furthermore, it is now believed that the primary wood, as well as the secondary, was able to increase in amount, because of continued meristematic activity in the parenchyma cells between the xylem tracheids. Growth processes, of course, cannot be studied directly in fossil material. They can only be deduced from a comparative study of a large number of specimens, and this has recently been carried out by Delevoryas.[50] He drew attention to a correlation between the size of the primary wood and that of the secondary wood and suggested that the only satisfactory explanation of this lay in the supposition that the primary wood at any particular level continued to increase with age. This being so, then the Medullosaceae are indeed a unique group.[168] Over forty species or varieties of the stem-genus *Medullosa* have been described, but the comparative studies of Delevoryas indicate that many of these represent no more than different stages of growth, or different levels, in the stems of a much smaller number of species.

Sutcliffia insignis (Fig. 12A) was first described in 1906,[154] from a roof nodule in a South Lancashire colliery and has recently been found in the U.S.A.[143] It is generally regarded as more primitive than *Medullosa*, and in some respects it occupies a position intermediate between the Lyginopteridaceae and the rest of the Medullosaceae. The stem attained a diameter of 12 cm and in the centre had a single large stele (1), very like that of *Heterangium*, consisting of a cylinder of manoxylic secondary wood surrounding an exarch protostele made up of mixed parenchyma and tracheids. From this central stele there arose smaller steles, varying in size and in number, branching and anastomosing in an irregular fashion, and eventually giving rise to leaf-traces. All these various steles retained the same organisation and even in the petioles the traces were still concentric, with a complete cylinder of secondary wood.

In some species of *Medullosa*, likewise, the leaf-traces retained their secondary wood for some distance and are, consequently, regarded by Delevoryas as being more primitive than those species in which the leaf-traces lacked secondary wood from their point of origin. The most primitive species in this respect was *M. primaeva*, described from an American coal-ball. Its stem was about 2 cm in diameter and contained up to twenty-three steles, of which about five were larger than the rest. As in many species of *Medullosa*, these showed some degree of 'endocentric' development, in that there was a greater thickness of secondary wood towards the centre of the stem. The leaf-traces were merely stelar branches which continued to anastomose as they went out into the leaf-base, 'a stem system caught in the act of becoming differentiated into a leaf'.[50]

Medullosa anglica was more advanced in that only some of the leaf-traces had secondary xylem. Furthermore, there was much less branching and anastomosis of the stelar system in the stem. The number of main steles was usually three (sometimes four) and they were surrounded by a periderm (indicated in Fig. 12C by a broken line), but there were, in addition, accessory strands, probably entirely secondary in origin, surrounded by their own periderm (2) and also occasional rings of periderm which lacked a stele altogether (7). Leaf-trace bundles had their origin from the outer sides of the three main steles, and divided repeatedly to form a large number of small collateral bundles which entered the decurrent leaf-bases. The petioles, when found detached, are placed in the form-genus *Myeloxylon*. They were large (4 cm or more across) and nearly half the diameter of the stem that bore them. Inside were as many as seventy or eighty bundles, giving the appearance of a monocotyledon stem. The frond associated with *M. anglica* was probably a species of *Alethopteris* (possibly *A. lonchitica*).

Several other species of *Medullosa* had a small number of steles, e.g. *M. centrofilis* (Fig. 12B), *M. Noei* and *M. Olseniae*, and their leaf-traces were completely without secondary xylem. According to Delevoryas, these represent the culmination of one line of evolution within the genus. The other line of evolution, which continued into the Permian, involved the tangential expansion and lateral fusion (phylogenetically) of outer rings of steles, and the reduction in size of the inner steles to give 'star rings'. Different stages in the first of these processes are seen in *M. Leuckartii* (Fig. 12D) and *M. stellata* (Fig. 12E), the fusion being complete in the latter, so as to produce a ring of primary wood with secondary wood, both inside and out. One such trunk was as much as 50 cm across with forty-three central steles, some of which were 2·5 cm across. Reduction of the central steles

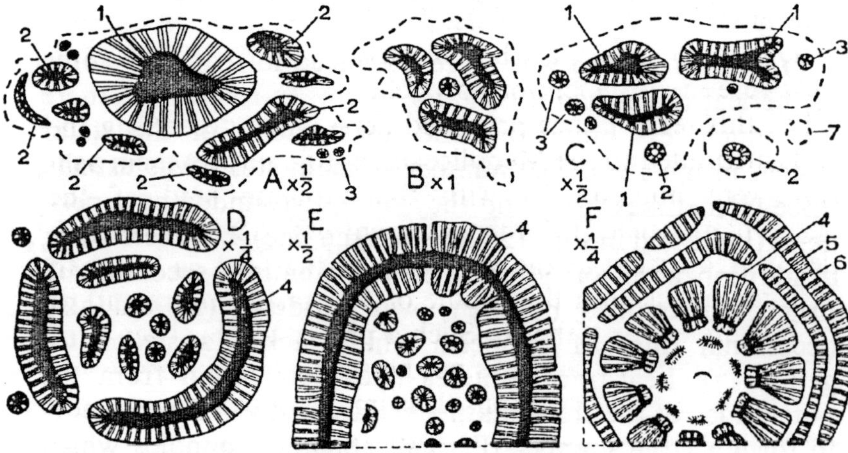

FIG. 12

Vascular systems of Medullosaceae

A, *Sutcliffia insignis.* B, *Medullosa centrofilis.* C, *M. anglica.* D, *M. Leuckartii.* E, *M. stellata.* F, *M. Solmsii* (var. *lignosa*). (Primary xylem—black; Periderm—interrupted; 1, main stele; 2, accessory stele; 3, leaf-trace)

(A, B, based on De Fraine; C, Scott; D, Stewart and Delevoryas; E, F, Weber and Sterzel)

can be seen in *M. Solmsii* var. *lignosa* (Fig. 12F). Relatively little tangential expansion of the outer steles had occurred in this species, but it illustrates another phenomenon—the development of rings of centrifugal secondary xylem (and secondary phloem) from successive anomalous cambia arising in the cortex.

At various times parallels have been drawn between the complex stelar arrangements of the Permian Medullosaceae and those of some present-day cycads, but, as Wesley[198] remarks, 'apart from anatomy, the reproductive organs of the two groups, especially the microsporangiate organs, are sufficiently different as to render the origin of the cycads from the Medullosaceae, at least the known members, as highly improbable'. Conversely, looking backwards, comparisons have sometimes been suggested with the polystelic Devonian ferns belonging to the Cladoxylales, but it seems very doubtful, in our present state of knowledge, that any phylogenetic relationship could be possible. The greatest similarities seem to lie with the Lower Carboniferous Lyginopteridaceae, and they were monostelic, not polystelic.

For this reason, it is proposed here, while describing the seeds attributed to the Medullosaceae, to interpret them as if they were homologous with the cupulate seed of the Lyginopteridaceae, i.e. the outer envelope, which is completely free except at the very base of the seed, is taken to be homologous with the cupule of *Sphaerostoma* instead of the integument as hitherto. Such a procedure has been proposed by Walton,[194] but has not been widely accepted yet as the basis of seed descriptions. It is important that the reader should be aware of this departure from custom, for, otherwise, he might have difficulty in reconciling the two different interpretations.

According to this new interpretation, Medullosan seeds differed in two important respects from those of the lyginopterids: 1. the integument (now to be called the inner integument) and the nucellus had become so intimately fused as to give the appearance of a vascularised nucellus; 2. the cupule had acquired a stony layer, thereby becoming more like an integument (it is proposed here to call it the outer integument, and not merely 'the integument', as is done by most writers).

Since 1828, when Brongniart first described *Trigonocarpus Parkinsonii*, there has been much confusion in the literature between the two generic names *Trigonocarpus* and *Pachytesta*, but the established practice at the present day is to use the former for specimens which are in the form of casts or impressions and the latter for petrified specimens that show internal structure. Accordingly, *Trigonocarpus Parkinsonii*, as described by Scott and Maslen,[155] is nowadays named *Pachytesta olivaeformis*. Nevertheless, the seeds of the Medullosaceae are usually grouped together under the name Trigonocarpales—a name which brings out the essentially trimerous nature of the seeds, for in most species the stony layer had three longitudinal ridges, representing the line of fusion of three sectors.

The largest species of *Pachytesta*, *P. incrassata*, attained an overall length of 11 cm and a diameter of 6 cm, but even the smallest, *P. hexangulata*, was large compared with lyginopterid seeds, for it was nearly 3 cm long. *P. olivaeformis* is one of the oldest of the eleven known species, being found in Lower Coal-measure deposits, while *P. illinoensis* is the youngest. The former was probably the seed of *Medullosa anglica* and the latter probably belonged to *M. Noei*. Both are illustrated in Fig. 13 in order to bring out the main trends in evolution within the genus.

Pachytesta olivaeformis (Fig. 13C) was about 4 cm long and 1·5 cm in diameter, cylindrical in the lower regions, but with the outer fleshy layer expanded into two flattened wings in the upper regions (Fig. 13D). The outer fleshy layer is usually poorly preserved, however, and there is just the possibility that some flattening may have occurred during fossilisation, but, in any case, it is true to say that the outer fleshy layer was thicker near the apex than at the base. Within it there ran six longitudinal vascular bundles. The stony layer round the body of the seed was about the size of an olive with three main ('commissural') ridges and six minor ridges, corresponding in position with the six vascular bundles, but, at the apex, the stony layer was extended into a long pointed beak surrounding the micropyle. The vascular supply to the inner integument consisted of a large number of longitudinal strands connected by transverse anastomoses into a network. The pollen-chamber was a simple bell-shaped structure with a central beak extending up into the micropyle.

Pachytesta illinoensis, at one time assigned to the genus *Rotodontiospermum*, was about the same size as *P. olivaeformis*, but it differed in several important respects. Perhaps the most noticeable feature is the longitudinal ridging of the stony layer, with vascular bundles alternating with the ridges (Fig. 13F). It is also important to notice that there were two series of bundles (seen in Fig. 13E) making a total of some forty-two in the outer fleshy layer. Separating the outer integument from the inner was a considerable gap and there was a well-developed stalk surrounded by a pendant skirt of tissue (3). These two features are shared by several other species and serve to strengthen the idea that the outer integument is homologous with the cupule of lyginopterid seeds. The vascular system of the inner integument consisted of up to twenty-four separate vascular bundles that stopped short at a point about two-thirds the way up. The pollen-chamber was dome-shaped, with a short beak and with a central conical plug of tissue.

In some species of *Pachytesta* the vascular system of the inner integument was tangentially expanded so that the bundles were almost overlapping, while in *P. Noei* it formed an almost continuous mantle, but, in all those species where the details are discernible, this innermost vascular system stopped below the level of the pollen-chamber, just as it does in present-day cycads.

Pachytesta hexangulata[165] is interesting in that a well-developed female prothallus was present, containing three archegonia beneath the pollen-chamber. Furthermore, there were pollen-grains in the pollen-chamber; they were about 365 μ across and corresponded in size and form fairly closely with those known to have been liberated from *Codonotheca*, one of the pollen-bearing organs believed, on other evidence, to have belonged to some member of the Medullosaceae. These pollen-grains are further interesting in that a multicellular male prothallus can be discerned within them.

Pollen-grains have also been seen in *P. vera*,[102] but these were much smaller, only 63 × 54 μ, and the cells of the male prothallus are less clearly visible.

Some five species of *Stephanospermum* are now known, all smaller than *Pachytesta*, for the largest was only 2 cm long, inclusive of a micropylar beak 5 mm long. The generic name refers to a circular crown at the apex of the body of the seed, but one species, *S. ovoides*, was almost without a crown. Instead, the apex of the seed had a flat saucer-like depression. Other features that have been used to distinguish *Stephanospermum* from *Pachytesta* are the presence of a continuous mantle of vascular tissue in the inner integument, and the lack of longitudinal ridges on the stony layer. However, *P. Noei* had an almost continuous vascular mantle, while *S. Stewartii* had only three to four rows of short stout tracheids, and *P. ovale* (= *Sarcospermum ovale*) lacked longitudinal ridges. It therefore seems that the two genera graded into one another.

Stephanospermum elongatum[84] is illustrated in longitudinal section in Fig. 13A. Its overall length was 3 cm, of which more than half was taken up by the beak, and the crown was about 2 mm high. Fig. 13B is of a transverse section at the level of the crown, and shows the four buttresses that crossed the trough and connected with the crown. Outside the stony layer there was a delicate papillate fleshy layer within which was a vascular system, but no particular pattern of vascular bundles can be discerned. The inner integument had a continuous mantle of tracheids that stopped short abruptly before reaching the pollen-chamber. Likewise, in all other species but *S. akenioides*, the vascular system stopped short of the pollen chamber. In this species Oliver[136] maintained that the vascular mantle extended right across the floor of the pollen-chamber. If so, this is probably the only seed in which this remarkable state of affairs exists, and it would seem that a critical re-examination should be undertaken.

In some specimens of *Stephanospermum* there was apparently no floor at all to the pollen-chamber, as illustrated in

Fig. 13A, but this presumably means that the seed became fossilised after this region had broken down. This is in accordance with the fact that pollen-grains with male prothallial tissue were present in the pollen-chamber. In some species, e.g. *S. caryoides* and *S. Stewartii*, where pollen-grains of a markedly different type have been found, it has been suggested that the pollen was extraneous, coming from the Cordaitales that formed the forests within which these medullosan plants were growing.

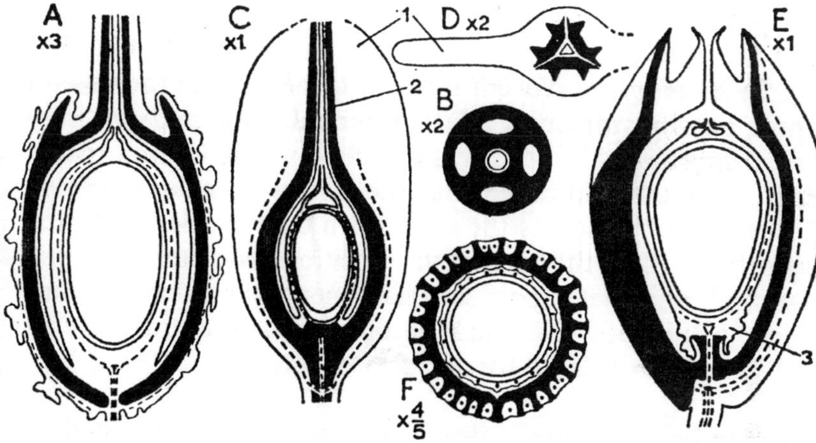

FIG. 13

Seeds of Medullosae

Stephanospermum elongatum: A, l.s.; B, t.s. through collar. *Pachytesta olivaeformis:* C, l.s. in the plane of the wings; D, t.s. through wing. *P. illinoense:* E, l.s.; F, t.s.
(1, wing; 2, beak; 3, skirt)
(A, B, after Hall; C, D, Scott, based on Maslen; E. F, Stewart)

Several other genera of seeds belonging to the Trigonocarpales may be mentioned briefly. Some differed internally in the type of decoration of the stony layer; thus, *Ptychotesta* had peltate projections, while *Hexapterospermum* had six equal, longitudinal ridges.[102] Others differed in their external form; thus, *Polylophospermum* had a peculiar cup-like extension of the outer integument round the micropyle, and

Codonospermum had a large basal air-chamber. Apart from the last of these, which was eight-angled and may not belong here at all, all had a basic trimerous symmetry and are, no doubt, correctly assigned to the Medullosaceae. Indeed, they tend to grade into one another, so that strict definition is difficult.

Several compression fossils have been found in which seeds corresponding in shape and size to members of the Trigonocarpales are attached to fronds. Sometimes the seed replaced the terminal pinnule of a pinna, as in *Neuropteris tenuifolia* and *N. heterophylla*. In *Pecopteris Pluckenetii* the seed was attached to the margin of a pinnule, at a vein ending. In *Alethopteris Norinii* it was attached to the rachis of a pinna, possibly replacing a lateral pinnule, but in *Emplectopteris triangularis* the seed was attached to the lamina of the frond. It should be noted that the last two species are from the Permian,[85] whereas the others are from earlier deposits, suggesting that the primitive position was terminal or marginal, and that superfical attachment was a derived condition, a 'phyletic slide' having occurred, much in the way that Bower postulated for the sori of ferns.

Pollen-bearing organs, too, have been found attached to Medullosan fronds. Thus, the synangium *Whittleseya media* was borne on a pinna of *Neuropteris Schlehanii*, whose frond form was identical with one that bore a seed closely resembling *Pachytesta vera*. In this species the synangium took the place of a lateral pinnule.[1] Most pollen-bearing organs, however, are known only as detached specimens, with the result that their attribution to the Medullosaceae depends on circumstantial evidence. Furthermore, most specimens are in the form of compressions, whose internal structure is not easily made out. However, Halle[86] devised a technique for softening, embedding and sectioning them, and it is to him, more than to anyone else, that we owe our present knowledge of these fascinating organs.

Whittleseya elegans (Fig. 11K) was shaped rather like a

wine-glass, about 5 cm high, made up of a large number of sporangia fused together side by side. *W. fertilis* was similar in shape, but was only 1·5 cm high. The pollen-grains inside the sporangia were elliptical, when flattened, and were large, ranging from 0·2 to 0·25 mm in length. For a long time these organs were described as pinnules with parallel venation, until it was shown that the 'veins' were actually sporangia. *Aulacotheca*, likewise, was misunderstood for some time, being identified as a seed belonging to the genus *Rhabdocarpus* by some workers, and to the genus *Holcospermum* by others. *A. elongata* (Figs. 11M and N) was fundamentally very similar to *Whittleseya*, except that it was closed at the apex. There were about nine sporangia fused into a club-shaped hollow synangium about 2 cm long. Its pollen-grains were similar in shape to those of *Whittleseya*, but were only 0·15 mm long. *Goldenbergia* (Fig. 11P), too, was thought at first to be a seed. The synangia were only about 8 mm long, made up of twelve to sixteen fused sporangia, and were borne on long stalks attached to one side of a naked branch-like pinna. The pollen-grains were very large indeed, being about 0·4 mm long. *Boulaya fertilis* was of similar shape externally, but it is believed that, instead of having discrete sporangia, there was a continuous zone of pollen-grains, as illustrated in the hypothetical transverse section Fig. 11Q.

Dolerotheca is one of the few genera to have been found in a petrified condition, which has made possible a much more detailed study of its internal structure. While Halle's interpretation, based on compressions, has been proved to be substantially correct, Schopf[152] has been able to put right some of the details. *D. formosa* must have looked remarkably like a half-orange, about 4 cm across (Fig. 11O). The inside consisted of a solid mass of elongated tubular sporangia, completely fused together and disposed in pairs in radial rows. The pollen-grains inside were up to 0·35 mm long and are identical with those found in the pollen-chamber of *Pachytesta illinoense*, from the same locality. *D. fertilis* was even larger, being about 5·5 cm across, and its pollen-grains, too, were larger (up to 0·46 mm). *D. Reedana* and *D. villosa* were both smaller, 2 cm and 1 cm across, respectively, but their internal organisation was the same.

Codonotheca (Fig. 11L) is interesting in that there was much less fusion of the adjacent sporangia than in the rest of the Whittleseyineae. Furthermore, it is represented in earlier deposits and may thus represent an ancestral type from which the rest might have evolved. *C. caduca* was up to 5 cm long, but, apart from its larger size, it is comparable in organisation with *Telangium* (Fig. 11D) and ultimately, therefore, with *Yarravia* (Fig. 11J).

Potoniea (Fig. 11R), while comparable in some respects with the Whittleseyineae, differed in several important characters, on the basis of which Halle favoured the creation of a separate group, the Potonieineae. *P. adiantiformis* consisted of a deep saucer, or wide cup, about 1·5 cm across, containing a large number of sporangia, which were completely free from each other. The pollen-grains inside them were much smaller than those of the Whittleseyineae, being only 70 μ long, and they differed also in being trilete instead of monolete. However, despite these differences, it seems fairly certain that *Potoniea* belonged to the Medullosaceae, for *Neuropteris Carpentieri* bore sporangia of a similar type, arranged in a similar manner.

Calamopityaceae*

Stems: *Stenomyelon, Calamopitys, Diichnia*

Petioles: *Kalymma*

As originally constituted, the Calamopityaceae contained just the one stem genus *Calamopitys*, which showed some similarities to the Lyginopteridaceae, but by 1936 the number had been increased to seven (*Stenomyelon*, *Calamopitys*, *Diichnia*, *Bilignea*, *Eristophyton*, *Endoxylon* and *Sphenoxylon*). Of these only the first three, however, were manoxylic. The last four were pycnoxylic. The group has always been rather ill-defined, or even 'nebulous',[110] and it is proposed here to follow Lacey[117] in transferring the pycnoxylic genera to the Cordaitales, thus restricting the Calamopityaceae to the three manoxylic genera. Even so, there was quite a wide morphological range, from solid protostele to a system of circum-medullary strands. Only stems and leaf-bases are known, so their true phylogenetic placing is necessarily uncertain. However, in those specimens where the cortex is preserved there was a well-developed 'Sparganum' system of fibres, and the size of the leaf-bases was such as to imply that the leaves were probably large and frond-like. It is on this evidence that the three genera are believed to belong to the pteridosperms.

Stenomyelon is represented by four species, three from the Lower Carboniferous of Scotland and one from the Upper Devonian of the U.S.A.[147] In transverse section the stem had a three-angled primary xylem with a strong resemblance to that of *Aneurophyton*. *S. primaevum*[125] was at a lower level of organisation than the other species because of the complete absence of parenchyma from its primary wood. In *S. Tuedianum*[110] the three lobes of xylem were separated by three narrow bands of paranchyma, while in *S. muratum* (the American species) and in *S. heterangioides*[125] there was a mixture of tracheids and parenchyma throughout the primary xylem. Surrounding the primary wood was a zone of secondary wood containing tracheids with multiseriate pits on the radial walls and wood-rays that ranged from one to six cells in width.

There appears to have been no protoxylem that could properly be described as cauline. Preparatory to the departure of a leaf-trace an exarch (or slightly mesarch) protoxylem appeared and then divided into two. The leaf-trace, single at first and accompanied for a short distance by secondary wood, also divided on its way out and then subdivided several times to give a large number of traces in the petiole, *Kalymma*. The fronds of *Stenomyelon* are believed by Long[125] to have been of the *Diplothmema* type with *Sphenopteridium* pinnules, and on the basis of frequent associations with stems and petioles he believes that the seeds of *S. Tuedianum* were *Lyrasperma scotica* (Fig. 10J).

Calamopitys americana, from Upper Devonian deposits of Kentucky, is known from stems that were up to 4 cm across. In the centre was a mixed pith, round which was a ring of mesarch strands, so numerous as to be almost contiguous. From one of these a leaf-trace branched off and, on its way out through the wood, split into two before dividing further in the cortex to provide a ring of bundles in the petiole (known as *Kalymma lirata*, when found detached from the parent axis). *C. annularis*, also from Kentucky, had very few tracheids mixed with the pith, while the German species *C. Saturnii* was even more advanced in having a pure pith, the number of circum-medullary bundles being reduced to six. *C. Foerstei*[147] is interesting because its leaf-trace origin was very similar to that in *Diichnia*. Two adjacent circum-medullary bundles were concerned in the origin of each leaf-trace, instead of just one. These gradually became confluent and then each gave off a single branch. The two branches then fused, on their way out, to give a single leaf-trace with two protoxylems.

Diichnia kentuckiensis,[146] also of Upper Devonian age, had a five-angled mixed pith with a single mesarch protoxylem strand in each angle. In the origin of the leaf-trace system two of these branched (but without approximating), each giving a trace and a reparatory strand. The two branches then subdivided further to give the numerous bundles that entered the leaf-base. Detached petioles known as *Kalymma resinosa* are believed to have belonged to *Diichnia*. The 'double-trace, unilacunar' type of node described for *Diichnia* was, at one time, thought to be very strange, but in recent years it has been recorded from a number of unrelated genera, e.g. *Ginkgo*,[81] *Ephedra*,[130] and it also occurs in some flowering plants, where Bailey[37] and his co-workers regard it as a primitive feature.

3

Glossopteridaceae and Mesozoic Pteridospermales

Glossopteridaceae*

Stems or Roots: *Vertebraria*

Leaves: *Glossopteris, Gangamopteris, Palaeovittaria, Rhabdotaenia*

Reproductive organs: *Scutum. Hirsutum, Lanceolatum, Ottokaria, Cistella, Pluma, Lidgettonia*

Whereas in Lower Carboniferous times the flora of the southern hemisphere was very much like that of the northern, by Upper Carboniferous and Lower Permian times a peculiar and completely different flora had come to occupy Australia, South Africa, South America, Antarctica and the Indian peninsula. These regions together make up the supposed continent of Gondwanaland, separated by the Tethys Sea from the other continents of the northern hemispere. This Gondwanaland flora is often called the Glossopteris Flora because of the widespread abundance of the leaves known as *Glossopteris.* These were tongue-shaped, as the name suggests (Fig. 14A), ranging in length from a few centimetres to several decimetres, and their venation was reticulate (Fig. 14C). They must have looked very much like the leaves of the modern fern *Polypodium musaeifolium* and, indeed, until recently they were widely believed to have belonged to ancient ferns.

Since they were first discovered, over a century ago, a large number of species have been described and, now that the structure of their cuticles has been examined, it seems clear that the genus should be split. Certainly, *Gangamopteris* was distinct, for it lacked the prominent midrib of *Glossopteris*. Two other genera are known which were tongue-shaped and may also have belonged to the Glossopteridaceae; these are *Palaeovittaria* and *Rhabdotaenia*. The former had a midrib in the proximal half of the leaf, but none in the distal half, and there was no anastomosis of the lateral veins. The latter had a strong midrib and forked lateral veins, between which anastomosis was extremely rare. However, as Surange and Srivastava[170] point out, the various genera grade into one another and cannot be sharply defined.

The probable appearance of the plant that bore these leaves has, for a long time, been a matter of debate, for only rarely have the leaves been found attached to a stem. There are indications, however,[178] that the leaves were borne in whorls on relatively delicate stems. Small scale-leaves with reticulate venation may well represent caducous bud-scales. *Vertebraria* is the name given to some very peculiar axes that are commonly associated with *Glossopteris* leaves, and which, for a long time, were thought to be the stems from which the leaves had fallen. One suggestion, however,[178] is that they represent rhizomes which bore leafy shoots, and another is that they represent roots. The xylem of *Vertebraria* was most unusual,[196] for it had a number of vertical radiating flanges, separated by very broad parenchymatous rays which were interrupted at intervals by horizontal plates of xylem. According to Sen,[157] the pitting of the tracheids was more like that of a member of the Coniferopsida than of the Cycadopsida, which casts some doubt on the suggestion that these stems do in fact belong to the Glossopteridaceae.

Many small seeds and microsporangia are to be found in association with glossopterid leaves, as a result of which it eventually became accepted that the group was gymnospermous, but until very recently no reproductive organs had ever been found actually in organic connection with the leaves. Great interest was, therefore, aroused when, in 1952, Plumstead[144] described two genera and six species of reproductive structures attached to the midrib of *Glossopteris* leaves. A few years later, further reproductive organs were described, bringing the number of genera to six, and of species to nineteen, some being found attached to *Gangamopteris* leaves and one to *Palaeovittaria*. Unfortunately, no petrified specimens have been found—only compressions, whose internal structure is not preserved—and Plumstead's interpretations have been criticised by other palaeobotanists. It is not even certain that these supposedly reproductive structures bore seeds at all.

With the exception of *Pluma*, all the reproductive bodies described by Plumstead are thought to have had a double structure, consisting of two appressed scale-like valves, or 'cupules', borne at the tip of a long pedicel, or sessile on the midrib of the leaf. The pedicellate forms were either in the axil of the leaf, as in *Ottokaria*, or were adnate in varying degrees to the midrib of the leaf.

The two valves of *Hirsutum Dutoitides* are illustrated in Figs. 14A and B. In A, the upper, protecting, valve is supposed to have been removed, so as to expose the inner face of the other valve, which is still attached to the leaf of *Glossopteris indica*. In the centre of this valve was a group of raised 'sacs', and round the margin was a fluted wing. At first the 'sacs' were assumed to be microsporangia, but further investigation showed that they contained much hard tissue, and it is now suggested that they were seeds, or that they contained seeds. The other valve, when young, was covered on its inner surface with hair-like bodies (hence the generic name), but, later on, these were shed and the inner surface of this valve appeared as in Fig. 14B. Early descriptions referred to this as the sterile valve, but in her later writings Plumstead[144] suggested that the hair-like bodies were probably microsporangia. *Hirsutum intermittens* was similar, but had a fairly long free pedicel.

Ten species of the genus *Scutum* have so far been described and, as the name suggests, they have shield-shaped valves. In one of them, *Scutum rubidgeum*, belonging to *Glossopteris tortuosa*, Plumstead claims to have seen the microsporangia to be large and flattened ('bract-like'). In *Lanceolatus*, of which two species were described by Plumstead, and a further two by Sen[156] from a coal-seam in India, the seed-bearing valve was completely fused with the leaf, only the upper valve being free. *Cistella* was similar, in that the ovuliferous valve was sunken into the tissue of the leaf, but, unlike *Lanceolatus*, it was not fused to the leaf, only the pedicel being adnate. In *Ottokaria*, belonging to *Gangamopteris* spp., the long-stalked reproductive organ was borne in the axil of the leaf. Figs. 14D and E illustrate the appearance of the inner face of the two valves, the ovuliferous valve, E, being surrounded by 'bracts', whereas the other valve, D, had a continuous wing surrounding it. *Pluma* was unlike all the rest and, indeed, were it not for the fact of its having been found attached to *Glossopteris* leaves it would certainly not have been included with them. It appears to have been a single, rather than a double, structure, with a characteristic drooping curved shape with a pendulous fringe, reminiscent of an ostrich-feather. Apparently, also they were either male or female, instead of being hermaphrodite as Plumstead claims for the other genera.

In 1958 Thomas[180] described yet another type of reproductive organ, *Lidgettonia africana*, attached to leaves of the *Glossopteris* type. These were stalked peltate bodies, about 5 mm in diameter, which, unlike Plumstead's genera, were not borne singly on the midrib of the leaf but scattered over the lower part of the lamina (Fig. 14F). Furthermore, the fertile leaves were much smaller than the sterile leaves. None of the peltate 'cupules' was shown to contain any sporangia or seeds, but the surrounding matrix of the rock contains abundant detached sporangia and many isolated seeds of uniform size and form. The sporangia closely resemble those described many years earlier by Arber[34] from Australian rocks containing abundant leaves of *Glossopteris Browniana*. The isolated seeds are comparable in form to those found elsewhere in rocks containing *Glossopteris* and *Gangamopteris* leaves, differing only in being slightly smaller.

Pant and his associates[140, 142] have recently carried out an extensive investigation of the many seeds and sporangia commonly found alongside leaves of the *Glossopteris* type, in Tanganyika, Australia and India. The African sporangia, *Arberiella*, were apparently borne terminally on slender, branched axes, and contained pollen-grains with two bladders ('bi-saccate'). The seeds had similar pollen-grains lodged in their micropyles. It would, therefore, seem fairly certain that the sporangia and seeds belonged to the same plant, and there is a high probability that both belonged to plants with the *Glossopteris* type of leaf.

The significance of the Glossopteridaceae is not yet fully understood, and is unlikely to be, until more is known of the detailed structure of their reproductive organs. Plumstead suggests that they may have been the ancestors of flowering plants, but this has been said, at one time or another, of every group of gymnosperms, and in this particular case would seem to be singularly premature and ill-advised. Thomas[180] even questioned whether all plants with leaves of the *Glossopteris* type should be classified together, for an examination of their cuticles suggests widely different affinities for different species. He concluded that 'this form of leaf may well have evolved by parallel or even convergent evolution'.

Peltaspermaceae*

Fronds: *Lepidopteris*

Seed-bearing organs: *Peltaspermum*

Pollen-bearing organs: *Antevsia*

This small family was established in 1933 by Thomas[176] on the basis of specimens of Triassic age collected by him in Natal, by Harris[88] in Greenland and by Antevs[31] in Sweden.

It is now known to have extended back into the Upper Permian and to have occurred in places as far apart as Argentina, Australia, Madagascar and China. Leaves, pollen-bearing and seed-bearing organs are known, but almost nothing is known of the stems that bore them or, therefore, of the general growth habit of the whole plant.

Five species of the frond genus *Lepidopteris* are recognised. They were lanceolate, up to 30 cm long, and bipinnate or tripinnate, with a single unforked rachis. The ultimate pinnules had a well-marked midrib and forking secondary veins. In all species there were a few pinnules ('zwischerfiedern') attached between the pinnae directly to the rachis, as illustrated in Fig. 14G. Stomata occurred on both the adaxial and abaxial surfaces ('amphistomatic'), but mostly on the abaxial surface; they were sunken in a pit, and were surrounded by five or six radially arranged subsidiary cells, as in most gymnosperms (but unlike those of the Bennettitales). In several species the rachis had a very uneven surface, because of raised lumps or blisters, and the occurrence of similar blisters on the axes of the reproductive organs provides good evidence that they belonged to one and the same plant.

The seed-bearing structure *Peltaspermum Thomasii*, associated with the frond *Lepidopteris stormbergensis*, was originally described under the name *Lepidopteris natalensis*,[176] and was illustrated (Fig. 14H) as having a number of stalked peltate heads, about 5 mm in diameter, arranged spirally around a main axis. However, Townrow[186] now believes that, instead of being peltate, they were bilaterally symmetrical, with a marginal stalk and that they were arranged pinnately on the main axis. The whole structure, therefore, according to him, is to be seen as a fertile frond, or sporophyll, and not as a branch bearing sporangiophores. In this particular species there were apparently only two seeds on each head (or, as Thomas remarked, perhaps it was that only two reached maturity), but in *P. rotula*, associated with *L. ottonis*, described by Harris from Greenland, each head was genuinely peltate and radially symmetrical, with ten to twelve seeds and ten to fifteen marginal lobes (Fig. 14I). The heads in this species were much larger than in *P. Thomasii*, being about 1·5 cm across. Its seeds were about 7 mm long and had a characteristic curved micropylar beak. On maceration, four distinct cutinised membranes can be demonstrated, corresponding respectively, it is thought, to (1) the outer cuticle of the integument, (2) the inner cuticle of the integument, (3) the cuticle of the nucellus and (4) the megaspore membrane. If this interpretation is correct, then it seems that the integument must have been free from the nucellus, as in Fig. 3H.

According to Townrow,[186] the pollen-bearing organs, *Antevsia*, should also be regarded as sporophylls, since they were bipinnate structures with alternate primary branching, even though the secondary branches were less regularly arranged (Fig. 14J). The ultimate branches of *A. Zeilleri* bore up to twelve pollen-sacs, about 2 mm long (Figs. 14K and L). These had a massive wall, showing a midrib and some stomata (surprisingly enough) on the dorsal surface, and a longitudinal line of dehiscence on the ventral surface. The presence of stomata is a most unusual feature in such an organ, and suggests that the wall of the pollen-sac must have been photosynthetic. The pollen-grains extracted from the pollen-sacs are shaped like a rugby football, with a single longitudinal furrow. They were completely without wings or bladders. A second species, *A. extans*, differed, in that the ultimate branches bore only four pollen-sacs. As Townrow points out, *Antevsia* is unlike any other known Mesozoic pollen-bearing organ. Although two species, *Sphenobaiera furcata* and *Antholithus Wettsteinii*, showed some approach to *Antevsia*, not only did they differ in shape and in the number of pollen-sacs, they also lacked stomata in the wall of the pollen-sac.

Corystospermaceae*

Fronds: *Dicroidium, Xylopteris*
Seed-bearing organs: *Umkomasia, Pilophorosperma, Spermatocodon*
Pollen-bearing organs: *Pteruchus, Pterorachis*

Like the previous family, this too was created by Thomas,[176] on the basis of specimens collected by him from Triassic rocks of Natal. Subsequently, a number of fronds and fertile structures have been discovered in Australia and Argentina[71, 72] and similar fronds unassociated with reproductive organs have been found in India.

The correct nomenclature of the fronds has been somewhat troublesome, particularly as regards their separation from *Thinnfeldia* and *Stenopteris*. However, a study of their cuticles has made the distinction more reliable and the fronds belonging to the Corystospermaceae are now placed in the two genera *Dicroidium* and *Xylopteris*.[185] In both genera the rachis forked near the base into two equal halves, as in many Palaeozoic pteridosperms. *D. odontopteroides* (Fig. 14R) was only about 10 cm long, was once pinnate and had open venation (Fig. 14S). *D. Feistmantelii* (placed by Frenguelli in the genus *Zuberia*), however, was much larger, up to 100 cm long, and was bipinnate. *Xylopteris* (Fig. 14M) had narrow linear pinnules, each with a single unbranched vein.

Eleven species of seed-bearing organs were found by Thomas, in association with fronds of *Xylopteris* and *Dicroidium*, and he placed them in the three genera *Umkomasia* (two species), *Pilophorosperma* (eight) and *Spermatocodon* (one). That they belonged to the same plants as the fronds is indicated by the similarity of their cuticles and stomata. In all species the seeds were borne singly in helmet-shaped cupules (Figs. 14 O and P), which in turn were borne at the tips of the ultimate branches of a pinnate structure (Fig. 14N). Thomas described the whole structure as an inflorescence, because of the small bracts and bracteoles that occurred at the nodes and elsewhere along the branches.

However, the branch system was markedly dorsiventral, the cuticle of the upper side being different from that of the lower side, and there is much to be said in favour of regarding the whole structure as a fertile frond. The cupule of *Umkomasia* was split into two halves, that of *Pilophorosperma* was lined with hairs, while that of *Spermatocodon* was bell-shaped, without any lining of hairs. It is perhaps surprising that stomata occurred on the inner surface of the cupules, as well as on the outer.

There was considerable variation in the shape of the seeds, but they all had a very characteristic long and bent bifid micropyle (Fig. 14Q). Maceration of isolated seeds gives evidence of pollination having occurred in many of them, and the pollen-grains inside the pollen-chamber were quite characteristic, with two large lateral wings, or bladders.

Identical pollen-grains can be extracted from the male organs *Pteruchus*, of which eight species were described by Thomas (however, Townrow[187] now reduces the number to three). *Pteruchus* was somewhat similar to the Palaeozoic *Crossotheca*. There was a central axis, up to 4 cm long, bearing short lateral branches generally arising in one plane. Each terminated in a peltate head (Fig. 14T), bearing on the underside more than thirty sporangia, up to 3 mm long, which dehisced by means of a longitudinal slit. According to Thomas these were probably bilocular, as in *Crossotheca*, but Townrow finds no evidence to support this suggestion. Thomas likened the sporangial heads to male flowers of *Populus*, and the structures bearing them to inflorescence axes, or catkins. However, Townrow, having shown that the sporangial heads had pinnate venation, suggests that they should be regarded as fertile pinnae, the structure bearing them being a fertile frond. Further support for this is given by the discovery of several specimens actually attached to a short length of the parent axis, without any sign of a subtending organ.

Pterorachis, described by Frenguelli[72] from Triassic rocks of Argentina, was similar, but was even more like a fertile

FIG. 14

Glossopteridaceae and Mesozoic pteridosperms

Glossopteridaceae. A, B, two halves of *Hirsutum* (= *Scutum*) *Dutoitides*, attached to leaf of *Glossopteris indica*; C, portion of leaf of *Glossopteris colpodes*, showing venation; D, E, two halves of *Ottokaria transvaalensis*; F, *Lidgettonia africana*, fertile leaf.

Peltaspermaceae. G, *Lepidopteris natalensis*, part of frond; H, part of seed-bearing structure, *Peltaspermum Thomasii*; I, reconstruction of cupulate disc of *P. rotula*, from which all but three seeds had fallen; J, *Antevsia* (*Antholithus*) *Zeilleri*, microsporophyll; K. L. opposite sides of microsporangiophore.

Corystospermaceae. M, *Xylopteris densifolia*; N, *Umkomasia Macleanii*; O, cupulate seed of *Pilophorospermum granulatum*; P, cupulate seed of *Pilophorospermum* sp.; Q, isolated seed; R, S, *Dicroidium odontopteroides*; T, microsporophyll, *Pteruchus stormbergensis*.

(A, B, D, E, after Plumstead; C, Pant; F–H, M–R, T, Thomas; I, Harris; J, based on photograph by Harris; K, L. S, after Townrow)

frond, for the rachis forked into two equal halves. It was found alongside fronds of *Dicroidium Feistmantelii* and some slender forking axes bearing paired cupules, like those of the South African genera.

Several other genera of forking fronds, besides *Dicroidium* and *Xylopteris*, have been found in Mesozoic deposits in various parts of the world, e.g. *Dichopteris Ptilozamites, Dicroidiopsis* and *Diplasiophyllum*. These may well, ultimately, prove to have belonged to pteridosperms, and even to the Corystospermaceae, but, until their association with reproductive organs has been established, their true affinities must remain in doubt.

Caytoniaceae*

Leaves: *Sagenopteris*

Seed-bearing organs: *Caytonia*

Pollen-bearing organs: *Caytonanthus*

Sagenopteris is the name given to a very characteristic type of leaf which has been known since 1828. Since then, leaves of this type, belonging to many species, have been found in rocks ranging from the Upper Triassic to the Lower Cretaceous. They had a slender petiole and, typically, four terminal leaflets, arranged in two pairs (Fig. 15A). Harris[98] has pointed out the curious fact that, not only was the whole leaf shed by means of an absciss-layer, but so also were the leaflets, a feature usually associated with dicotyledons. The leaflets (Fig. 15B) had a well-marked midrib and arched, forking lateral veins with lateral connections, giving an anastomosing system like that of *Glossopteris*.

Despite the fact that these leaves had been known for such a long time, nothing was known until 1925 of the rest of the plant that bore them. Even now, hardly anything is known of its stem, except for a tiny branched twig bearing some bud-scales of *Sagenopteris Phillipsii*.[92] This scarcely tells us anything of the growth habit of the plant, except that, as Harris remarks, at least we know that the leaves were not borne directly on a stout trunk like that of a cycad or a tree-fern.

The reproductive organs were first described by Thomas[175] in 1925, from Mid-Jurassic rocks of Yorkshire, and consisted of seed-bearing and pollen-bearing organs. Then, in 1929, Edwards[59] described some seed-bearing organs associated with *Sagenopteris Goeppertiana* from Sardinia, and in 1933 Harris[90] described both types of reproductive structure from Lower Jurassic rocks of Greenland.

Of the seed-bearing organs, Thomas described two genera, *Caytonia* and *Gristhorpia*, but subsequent workers have merged these into the one genus *Caytonia*. The seeds were borne in small fruit-like structures, rather like currants, which in turn were borne on a rachis some 5 cm long (Fig. 15C). In its epidermal structure this rachis was strongly dorsiventral, which lends support to its interpretation as a megasporophyll. There is no convenient term by which to call these fruit-like bodies, however. The term 'cupule' is applied to them, but the seeds were almost completely enclosed within them. Perhaps the best course is to follow Harris[98] in calling them 'fruits', using the word, not in its formal morphological sense, but in its normal everyday sense for a juicy pulp, surrounding seeds. Fig. 15D illustrates the appearance of such a fruit, the outline of the enclosed seeds being visible in the compressed form in which it was preserved. In *C. Sewardii* there were about eight seeds in each fruit, in a single arched row, in *C. Nathorstii* about fifteen, and in *C. Thomasii* about thirty in a double row. At one side of the stalk of the fruit, a lip or flange can be seen (I), bearing about the same number of ridges as there were seeds inside. This was thought by Thomas to be a kind of stigma, on which the pollen-grains germinated, but it is now known that the pollen-grains reached the micropyle of the seeds. Accordingly, Harris interprets the fruit (Fig. 15E) as having had a separate channel leading from the flange to each seed, and supposes that there was a 'drop-mechanism', as in many living gymnosperms, the pollen-grains, once trapped in the drop, floating up the channels to the seeds.

In this it is possible that the two lateral bladders on each pollen-grain may have acted as flotation devices.

Isolated seeds occur abundantly in the rocks, along with the fruits, and were about 2 mm × 1 mm. They can be sectioned with a microtome, but only after softening by means of a long and tedious process, involving the use of boiling alcoholic potash for some five weeks, followed by treatment with hydrofluoric acid, and then embedding in celloidin. It then becomes clear that the single integument was free from the nucellus. Indeed, the seed was a typical gymnosperm seed, except in one particular. This was the extraordinary thickness of the cuticle covering the nucellus. There is no evidence of any vascular system in the integument.

The pollen-bearing organs, *Caytonanthus*, were figured as early as 1829, but were described as leaves, and again in 1919,

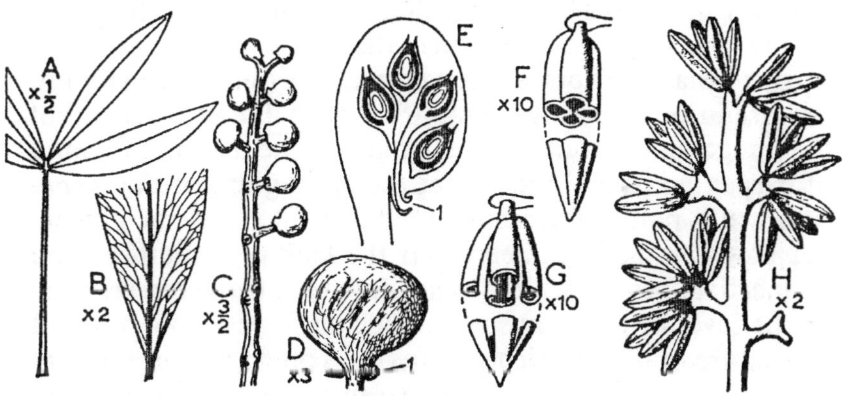

FIG. 15

Caytoniaceae

A, *Sagenopteris Phillipsii*, leaf with four leaflets. B, *S. colpodes*, showing reticulate venation of leaflet. C, *Caytonia Nathorstii*, fruiting rachis, from which the lower 'fruits' have been shed. D, *C. Sewardii*, external appearance of 'fruit'. E, *Caytonia* 'fruit' in l.s., as interpreted by Harris. F. G. *Caytonanthus Arberi*, restoration of synangia, cut across so as to show the loculi (F, before and G, after dehiscence). H, *Caytonanthus Arberi*, male sporophyll.

(1, flange, or lip, of 'fruit')

(A–C, E–G, after Harris; D, H, Thomas)

under the name *Antholithus*, when they were supposed to belong to some member of the Ginkgoales. However, the work of Thomas, and of Harris, has made it clear that in structure *Caytonanthus* was quite unlike any other microsporophyll. There was a dorsiventral rachis (with the upper epidermis different from the lower) bearing opposite, or sub-opposite, pinnae. The pinnae branched irregularly, and each terminal branchlet bore a single synangium. Fig. 15H, taken from the early account by Thomas, shows the general appearance of the whole structure, but does not show the ultimate branch-tips. One of these is illustrated in Fig. 15F, with its pendulous synangium cut across so as to show the four loculi. On dehiscence, the four pollen-sacs separated from each other, except at the tip, where they still remained joined together (Fig. 15G).[98]

Few fossils have created such a stir among morphologists as did the Caytoniaceae when they were first discovered. This is because Thomas believed that they provided clues to the origin of flowering plants. Not only did he believe that the flange on the fruit was a stigma and the fruit itself a kind of carpel, he believed also that the synangium of *Caytonanthus* was a point of agreement with the angiosperm stamen, despite its radial symmetry and the lack of a filament and connective. However, it is clear now that *Caytonia* was still at the level of a gymnosperm and, until we know more of the nature of its fruit, decisions as to its phylogenetic relationships should be withheld. The same may be said of the Corystospermaceae, for here too the nature of the cupule is not fully understood. Some botanists have suggested that both groups should be separated from the pteridosperms and given equivalent status, but, as Townrow points out, so long as we define the pteridosperms as 'gymnospermous plants with leaves, pollen- and seed-bearing organs pinnate; reproductive structures not aggregated into cones or flowers', then there is no reason why the Corystospermaceae, and, indeed, the Caytoniaceae too, should not be retained within the Pteridospermales.

4

Bennettitales (=Cycadeoideales)

Stems with wide pith, stout and pachycaulic, or relatively slender and forking. Leaves compound (rarely simple) with open (rarely closed) venation. Stomata syndetocheilic. Reproductive organs in hermaphrodite or uni-sexual 'flowers', protected by numerous bracts. Ovules stalked, very numerous, scattered over a conical, cylindrical or dome-shaped receptacle, along with interseminal scales, more or less united at the distal end to form a shield, through which the micropyles protruded. Seeds with two cotyledons. Pollen-bearing organs in a whorl, free or united, pinnate or entire, with numerous microsporangia, usually in 'capsules'.

1. **Williamsoniaceae***

   Williamsonia (whole plants, male and female flowers)

   Fronds: *Ptilophyllum, Otozamites, Dictyozamites, Pterophyllum*

   Seed-bearing organs: *Bennetticarpus, Vardekloeftia, Westersheimia*

   Pollen-bearing organs: *Bennettistemon*

   Hermaphrodite flowers: *Sturiella*

2. **Wielandiellaceae***

   Wielandiella (fronds=*Anomozamites*)

   Williamsoniella (fronds = *Nilssoniopteris*)

3. Cycadeoideaceae*

Cycadeoidea (= *Bennettites*)

Of all the various groups of fossil gymnosperms, the Bennettitales, or Cycadeoideales, have aroused perhaps the widest interest and, considering the relatively small number of genera that are known, the literature dealing with the group is voluminous. Interest was first aroused because the best-known genus *Cycadeoidea* had reproductive organs that were superficially like the flower of an angiosperm, with an elongated receptacle bearing perianth-like bracts and a whorl of pollen-bearing microsporophylls below the ovuliferous region. These formed the basis of a theory of angiosperm evolution that achieved wide acceptance for a time,[35] until it was realised that the similarities were only superficial. However, interest in the group was maintained because of the peculiar way in which the ovules were borne, often on long stalks, and because of the problems of interpretation of the morphological nature of the intersimenal scales scattered between the ovules. Then, too, there has been the question of a possible relationship to the Cycadales. Some of the later members of the group must, indeed, have looked superficially very similar to *Cycas*, with its stout stumpy trunk and a crown of pinnate leaves, but, as will be seen in a future chapter, this similarity, too, is only in external appearance. Not only were their reproductive organs quite unlike those of cycads, but so also were their stem anatomy and their stomatal structure.

Until 1913 no one had been able to distinguish with certainty between the leaves of the Bennettitales and those of the early cycads. So numerous were such leaves in the Mesozoic that this period was often called 'the age of cycads'. In that year Thomas and Bancroft[181] demonstrated that if cuticle preparations are made the two kinds of leaf can readily be distinguished by looking at their stomata. Bennettitalean stomata have two equatorial subsidiary cells, which are the same length as the guard-cells, an arrangement that also occurs in the living genus *Welwitschia* (to be described in a later chapter). Observation of the early stages of ontogeny of the stomata in this genus shows that both the subsidiary cells and the guard-cells are formed from a single mother-cell, a process which is called 'syndetocheilic', and it is assumed that similar processes occurred in the Bennettitales. By contrast, the stomata of cycads, and of most other plants, are called 'haplocheilic', because the stomatal initial cell provides only the two guard-cells, any subsidiary cells being formed by modification of adjacent epidermal cells. These observations of Thomas and Bancroft were confirmed by Florin,[61] and it became clear that, although cycads were indeed in existence by Mesozoic times, the great majority of cycad-like fronds of that age belonged to members of the Bennettitales.

The name *Williamsonia* was first used in 1870[200] for a combination of foliage and reproductive organs, referred to the species *W. gigas*, from Jurassic deposits in Yorkshire, but, since then, examples have been found in rocks as old as the Triassic. Williamson's reconstruction of the whole plant, which he called *Zamia gigas*, believing it to be a cycad, showed it as having a slender trunk, about 2 mm tall, like that of a tree-fern, with a crown of pinnate leaves and some flower-like buds on slender stalks. In 1932 Sahni[149] published a reconstruction of *W. Sewardiana*, based on petrified material from the Upper Gondwana beds of India (Fig. 16F). Trunks of this kind, previously known as *Bucklandia*, were up to 2 m tall, and occasionally showed unequal branching.

The leaves associated with *W. Sewardiana* were of the kind called *Ptilophyllum*. Like most bennettitalean leaves, those belonging to this form-genus were once-pinnate, the distinguishing feature being the attachment of the lateral pinnae by a broad base to the upper side of the rachis, which they almost covered (Fig. 16A). *Pterophyllum* (Fig. 16B) was similar, except that the pinnae were attached to the sides of the rachis. *Zamites* (Fig. 16C) differed from *Ptilophyllum* in

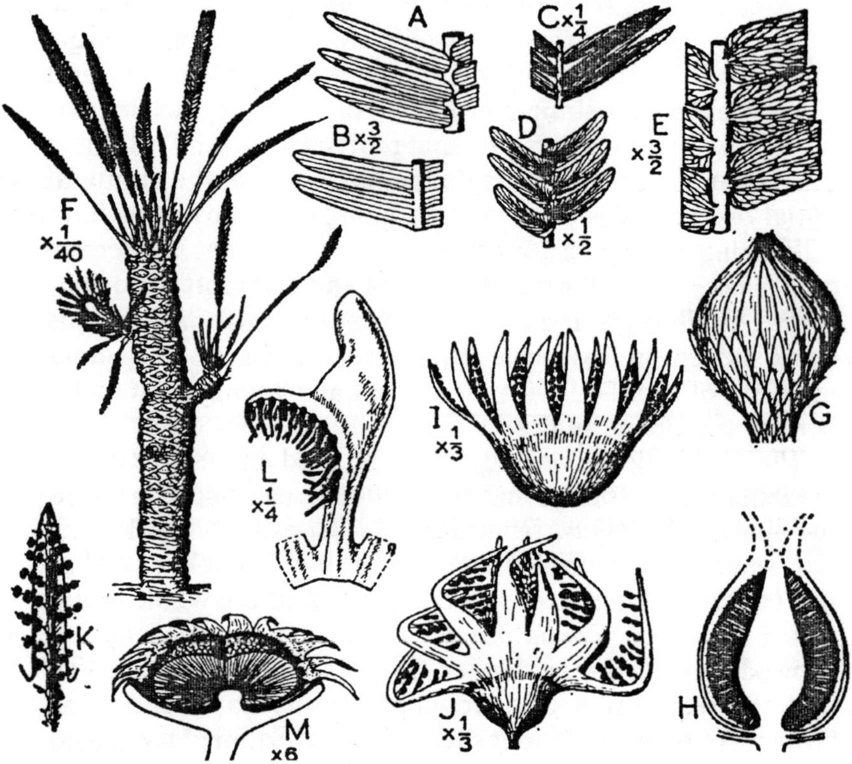

FIG. 16

Williamsoniaceae

Pinnules: A, *Ptilophyllum Anderssonii*; B, *Pterophyllum Thomasii*; C, *Zamites gigas*; D, *Otozamites Bechei*; E, *Dictyozamites falcatus Williamsonia:* F, reconstruction of *W. Sewardiana*; G, flower-bud of *W. gigas*; H, l.s. ovuliferous flower of *W. gigas*; I, pollen-bearing organ of *W. whitbiensis*; J, pollen-bearing organ of *W. spectabilis*; K, upper side of one lobe of the pollen-bearing organ of *W. spectabilis*, showing pinnules with pollen-capsules; L, one lobe of the pollen-bearing organ of *W. santalensis*; M, *Sturiella* (*Sturianthus*) *Langeri*, hermaphrodite flower.

(A, after Halle; B, Harris; C–E, Seward; F, Sahni; G, Williamson; I, K, N, Nathorst; J, Thomas; L, Sitholey and Bose; M, based on Kräusel)

that the pinnae were constricted at the base. The venation in all these genera was parallel, or slightly diverging, with occasional dichotomies, but that of *Otozamites* (Fig. 16D) was spreading. *Otozamites* was also characterised by the arcuate expansions at the base of the pinnae. Only one genus had anastomosing veins: this was *Dictyozamites* (Fig. 16E) which, otherwise, resembled *Otozamites*.

In Sahni's reconstruction of *W. Sewardiana* the left-hand branch is shown as terminating in an ovuliferous 'flower'. Within a series of perianth-like bracts was a dome-shaped receptacle covered with long-stalked ovules and club-shaped interseminal scales. Some morphologists describe such reproductive organs as 'strobili', or as 'cones', but neither term seems appropriate. Likewise, some may argue that the term 'flower' is inappropriate and that it should be reserved for the angiosperms. However, it is proposed here to call them flowers, as a term of convenience, so long as it is clearly understood that no direct homology with the flower of an angiosperm is implied by so doing. Furthermore, it is proposed to follow the custom of floral morphologists in calling them male and female, instead of pollen-bearing and ovuliferous, fully realising that these are terms that, strictly, should be reserved for gametophytes.

An unopened female flower of *Williamsonia gigas* is illustrated in Fig. 16G. Exactly what it was like inside, however, is somewhat doubtful, for the specimens examined by Williamson were in the form of casts. There seems no doubt that there was a central receptacle, bearing crowded stalked ovules and interseminal scales (Fig. 16H), but there has been some disagreement as to the apex of this receptacle. Some specimens suggest that it may have been expanded into a funnel-shaped corona, but this may be an artifact, caused by the pressure of the bracts against the matrix. In any case, however, it seems that the ovules and interseminal scales did not extend right over the apex of the receptacle.

Harris[89] has described several species of female flowers that occur, along with *Lepidopteris*, in Greenland. Rather than put them in the genus *Williamsonia*, he established a non-committal form-genus *Bennetticarpus*, to which he suggested that most bennettitalean female flowers ought to be transferred. *B. exiguus* was relatively small, being only 1·5 cm long and 1 cm across, but *B. tylotus* was as much as 4 cm in diameter and spherical in shape. Maceration removed everything except the cutinised outer surfaces of the interseminal scales, cutinised micropyles and various layers of the seed, representing the outer surface of the nucellus and the inner surface of the integument (Fig. 17I illustrates a similar arrangement in *Wielandiella*).

Concerning the morphological interpretation of these interseminal scales, Seward[20] suggested a homology with the stalked ovules, and Harris has adduced further evidence to support this hypothesis.[89] It was provided by *Bennetticarpus crossospermus* (Fig. 17H) in which each micropyle projected through a hole in the centre of a cutinised plate, which he regards as the outer surface of a cupule. Not only was this plate similar in size and shape to the interseminal plates, but in addition each interseminal plate had a central raised boss which accentuated the similarity, corresponding as it did to the projecting micropyle of the ovule. The similarity was even greater when the ovules were immature. Accordingly, Harris regards the interseminal scales as having 'formed by the diverted development of seed initials'. The suggestion that *B. crossospermus* had a cupule is interesting, for it is not visible in later species. It was, however, present in *Vardekloeftia*. The 'fruit' of *V. sulcata*, from the *Lepidopteris* beds of Greenland, was spherical, up to 3 cm across, and was borne on a slender stalk. Unlike *Bennetticarpus*, it had only a small number of seeds (two to six), which were relatively large, 7 mm long, and which projected beyond the interseminal scales. The presence of a distinct cupule seems to have been a primitive feature within the group, but whether in more advanced forms it became fused with the integument, or whether it just failed to develop, is not known.

Many detached male flowers are known, some described under the generic name *Williamsonia*, while others are placed in the form-genus *Bennettistemon* (equivalent in status to *Bennetticarpus*). They showed a much greater range of form than did the female flowers, but all consisted basically of a whorl of sporophylls which were more or less united into a cup. In some species, the sporophylls branched pinnately, e.g. *W. spectabilis* (Figs. 16J and K), but in most they were unbranched, e.g. *W. whitbiensis* (Fig. 16I). In *W. santalensis*, recently described from the Upper Gondwana of India,[162] each sporophyll was bifid, one half being sterile and the other fertile. This interpretation of the specimen is, however, only a tentative one, and it is possible that the sterile portion was really abaxial and that, on compression, it came to look as if it were lateral. Each sporophyll was about 10 cm long, and the whole flower, when fully expanded, must have been as much as 26 cm across.

The fertile region of *Williamsonia santalensis* bore two rows of finger-like appendages, 2 cm long, with two series of small chambers inside. Most of the other male flowers, however, bore their pollen in purse-like 'capsules', which some morphologists have described as 'synangia', but which others regard as equivalent to infolded pinnules. Evidence for the latter interpretation is provided by *Bennettistemon ovatum*, a small ovate leaf-like structure, 7 mm long, with separate sporangia on the upper surface. Harris suggests that it might have been a pinnule belonging to a compound microsporophyll. Fig. 16K shows diagrammatically the way in which the capsules were borne, in two rows, along the ultimate branches of the pinnate sporophyll of *Williamsonia spectabilis*, while Fig. 16I illustrates their superficial arrangement on the adaxial surface of *Williamsonia whitbiensis*.

It is fairly certain that most of the flowers so far described were unisexual, although for a time there were suggestions that the whorl of microsporophylls might have been borne beneath the ovuliferous receptacle (gynoecium). Such an arrangement did, indeed, occur in *Sturiella Langeri*,[113] from the Triassic of Austria, whose hermaphrodite flower is illustrated in Fig. 16M.

Members of the Wielandiellaceae differed from the rest of the Bennettitales in having relatively slender forking stems. Nathorst's reconstruction of *Wielandiella angustifolia*[134] (Fig. 17E) illustrates the way in which the female flowers were borne in the angles of the branches. The central club-shaped receptacle (Fig. 17G) was covered with ovules and interseminal scales, except for a tiny region at the apex. The leaf *Anomozamites Nilssonii* (Fig. 17F) was once-pinnate, with pinnae of irregular length, but usually less than twice as long as broad. Little is known of the male flowers because of their poor preservation.

The other genus of the *Wielandiellaceae*, *Williamsoniella*, of which four species are known from the Jurassic of Yorkshire, had hermaphrodite flowers, and was unique in that its leaves, *Nilssoniopteris*, had a simple lamina. The reconstruction prepared by Thomas[174] of *W. coronata* (Fig. 17A) shows the flowers as being in the angles of the branches, but Zimmermann[201] has figured a specimen from the Geological Institute at Tübingen in which the flower occurred in the axil of a leaf. The half-flower, as interpreted by Thomas (Fig. 17B), shows a number of microsporophylls in a whorl, beneath the gynoecium. According to him, the microsporophylls were fleshy, with the pollen-capsules sunken into them, but Harris[96] has reinterpreted them as pinnately lobed (Fig. 17D). He also demonstrated that the young flowers were enclosed in hairy bracts. The apex of the gynoecium was sterile (Fig. 17C), and the young microsporophylls were pressed tightly against it. The fertile region had closely packed interseminal scales, between which the micropyles of the seeds projected, and maceration showed that the integument of the seed was free from the nucellus (Fig. 17I).

The principal genus (and some say the only genus) of the

Cycadeoideaceae is *Cycadeoidea*, of which more than thirty species are known, ranging in age from Upper Jurassic to Upper Cretaceous. Most of the specimens come from American sources, but many have also been found in the Isle of Wight and Portland, as well as in Europe, Russia and India. They occur as petrified trunks, of massive proportions,

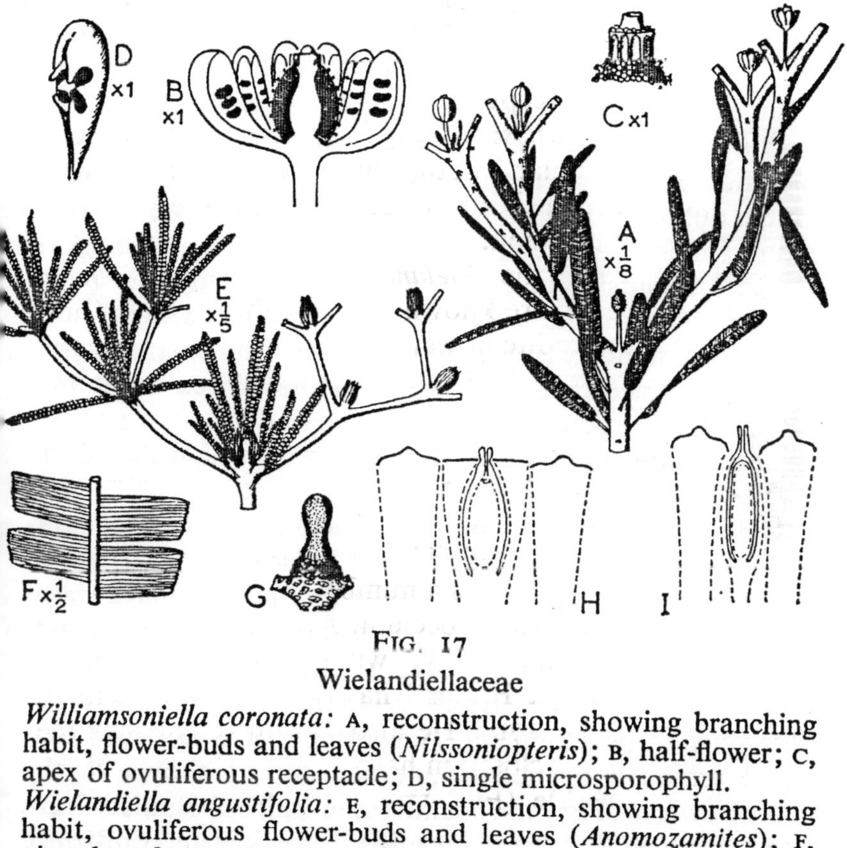

FIG. 17

Wielandiellaceae

Williamsoniella coronata: A, reconstruction, showing branching habit, flower-buds and leaves (*Nilssoniopteris*); B, half-flower; C, apex of ovuliferous receptacle; D, single microsporophyll.

Wielandiella angustifolia: E, reconstruction, showing branching habit, ovuliferous flower-buds and leaves (*Anomozamites*); F, pinnules of *Anomozamites Nilssonii*; G, club-shaped ovuliferous receptacle.

H, I: diagrams for comparison of l.s. seeds and interseminal scales in *Bennetticarpus crossospermus* (H) and *Wielandiella angustifolia* (I).

(A–C, after Thomas; D, F, H, I, Harris; E, G, Nathorst)

up to 60 cm across, but usually less than 1 m tall. Some were unbranched, but others branched near the ground into a cluster of short trunks.

In the centre of the trunk was a large pith surrounded by a ring of endarch primary wood and a narrow zone of secondary wood, composed of scalariform tracheids. The wood was interrupted by numerous broad rays, associated with the departure, in a spiral sequence, of leaf-traces (Fig. 18D), but, in addition, there were many uniseriate, or biseriate, rays. Each leaf-trace was single at the point of origin, and C-shaped, but, as it passed through the cortex, it split up into a number of mesarch strands arranged in the shape of a horseshoe. It is important to note that the leaf-traces passed straight out to the leaf, without circling round in the cortex, a feature which marks these trunks off from those of modern cycads. The outside of the trunk was protected by a heavy armour of persistent leaf-bases. No specimens have been found with mature leaves actually attached to the trunk, but it is assumed that the leaves must have been borne in a crown, at the apex of the trunk. That they were simply-pinnate is indicated by a study of buds, in which the young leaves were still unexpanded.

Flower-buds are frequently visible in the axils of leaf-bases, and in several species, e.g. *Cycadeoidea Dartonii* (= *Monanthesia Dartonii*), as many as 500 have been observed on a single trunk, one in the axil of every leaf-base. A remarkable feature of several species is that all the flower-buds on any one trunk are found to have been at the same stage of development. Accordingly, it is assumed that they must have opened simultaneously, as shown in the reconstruction (Fig. 18A), and some palaeobotanists have even concluded that the plant might have flowered only once during its lifetime.

The vast majority of species had hermaphrodite flowers, borne on a short pedicel and protected by as many as a hundred hairy bracts. The pollen-bearing region (the 'androecium'), as reconstructed by Wieland,[24] and illustrated in Fig. 18B, has been generally accepted and is widely reproduced in textbooks of palaeobotany. According to this interpretation, it consisted of a whorl of about twenty pinnate sporophylls, more or less united at the base, and bearing bean-shaped pollen-capsules in two rows along each pinna. The left-hand side of Fig. 18B shows them in the young state, folded in round the central gynoecium, and the right-hand half shows them as Wieland believed they would have looked when expanded. However, it must be emphasised that no flowers have actually been found in the expanded condition, and a recent re-examination by Delevoryas[53] suggests that the structure of the androecium was such that unfolding would have been impossible. Wieland himself had observed that in *Cycadeoidea colossalis* there was a dome-shaped dorsal extension, and that *C. ingens* probably had a similar structure, no longer visible because erosion had removed the distal part of the flower-buds. Delevoryas has confirmed the presence of such dorsal extensions, as illustrated in Fig. 18C. He suggests that the sporophylls could be likened to the segments of an orange. Although free in the distal regions, they were fused to each other lower down. Within each there were synangia (shown in black in Fig. 18C) hanging from trabeculae which extended from the outer wall to the inner. A line of dehiscence is visible round the base, suggesting that the androecium might have been shed as a single unit. Self-pollination might have occurred before it became detached and then, as it rolled along the ground, further pollen might have been shaken out and blown by the wind, thereby enabling some degree of cross-pollination to occur. Another possibility suggested by Crepet,[208] is that insects with chewing mouth-parts might have played a part in pollination.

The bean-shaped pollen-capsules (Figs. 18F and G) were about 3·5 × 2·5 mm and contained between twenty and thirty longitudinal pollen-sacs. The wall of the capsule was made up of palisade-like cells and dehisced by means of an apical slit, which allowed the capsule to open into two valves.

The gynoecium consisted of a conical receptacle (or a spherical one in *C. Gibsonianus*—Fig. 18H—and in similar unisexual flowers), bearing hundreds of tiny stalked ovules, only about 1 mm long, and about an equal number of interseminal scales. Maceration gives a cuticle preparation, on the basis of which Harris[88] interprets the seed as illustrated in Fig. 18E, the integument being fused to the nucleus,

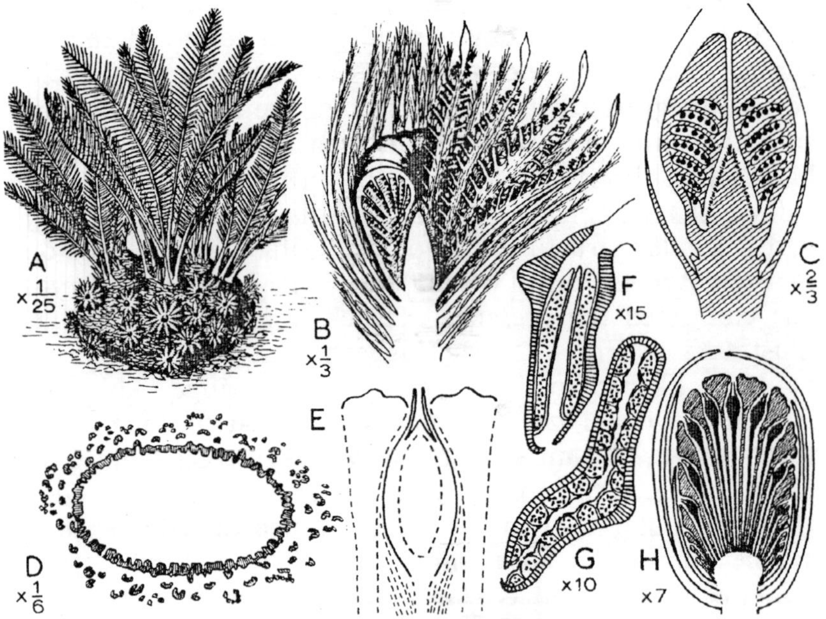

FIG. 18

Cycadeoideaceae

A, reconstruction of *Cycadeoidea Marshiana*. B, reconstruction of a single flower of *Cycadeoidea* (cf. *ingens* or *dacotensis*), as interpreted by Wieland, left side unexpanded, right side fully open; C, reconstruction as interpreted by Delevoryas. D, stele of *C. Saxbyanus*. E, l.s. seed and interseminal scales of *Cycadeoidea* sp. F, G, l.s. and t.s., respectively, of pollen-capsule of *C. dacotensis*. H, l.s. female flower of *C. Gibsonianus*, showing stalked seeds and interseminal scales (diagrammatic).

(A, based on Hirmer; B, Wieland; C, Delevoryas; D, after Carruthers; E, Harris; F, G, Wieland; H, based on Scott)

except at the apex. There was, however, an outer covering to the seed, and its stalk, consisting of elongated tubular cells, which Harris believes might represent a vestige of the cupule that occurred in the earliest members of the Bennettitales.

Before leaving this fascinating group, one ought to consider briefly their possible ancestry. Were it not for their stalked ovules there would be little difficulty in imagining that the Bennettitales were descended from the Pteridospermales, for, although there is some doubt as to the exact nature of the androecium of *Cycadeoidea*, that of earlier members of the group is easily interpreted as basically frond-like. However, it stretches the imagination to the limit to suppose that stalked ovules and interseminal scales are homologous with fronds. Harris points out that we need to know more about their phyllotactic arrangement and about their vascular supply. In the absence of such knowledge it would almost seem necessary to treat the group as unrelated to the pteridosperms, but there is one fossil plant which does provide some slight evidence to the contrary. This is *Westersheimia*, described by Kräusel,[114] from the Triassic of Austria, where it occurs along with the earliest species of *Bennetticarpus* (*B. Wettsteinii*). In this unique plant the ovules and interseminal scales were borne on the ultimate segments of a pinnate structure. However, even if this early member of the group does indeed represent a link with the pteridosperms, there still remains much to be discovered about the way in which the rest of the group evolved from it.

5

Pentoxylales

Fossil plants, habit unknown, but probably shrubs or very small trees. Long and short shoots, the latter bearing reproductive organs terminally, and spirally arranged leaves. Stems polystelic. Wood-rays uniseriate. Leaves thick, simple, lanceolate. Venation open (anastomoses very rare). Female organs like stalked mulberries; seeds sessile, united by fleshy outer layer of integument. Male organs consisting of a whorl of branched sporangiophores, fused basally into a disc.

Pentoxylaceae*

Stems: *Pentoxylon, Nipanioxylon*

Leaves: *Nipaniophyllum*

Seed-bearing organs: *Carnoconites*

Pollen-bearing organs: *Sahnia*

This small but remarkable group is of relatively recent discovery, the name Pentoxyleae having been proposed in 1948 by Sahni[150] in a paper that summarised the work of Srivastava, carried out during the previous decade, on stems, leaves and seed-bearing organs. The male organs were unknown until Vishnu-Mittre[191] described them in 1953. As this latter author points out, the group shows some features in common with the Bennettitales, some with the Cycadales, and some features that are unique. Accordingly, most palaeobotanists agree with Lam[118] that the group should be given a status equivalent to that of the Bennettitales and the Cycadales.

Fossil remains of the Pentoxylales are of Jurassic age. They are known from one locality in north-eastern India (where they occur as petrifactions in silica) and have recently been discovered in New Zealand, too.[99b] Two genera of stems are recognised, *Pentoxylon* and *Nipanioxylon*, both of which are remarkable for their polystelic structure. *Pentoxylon Sahnii* ranged in size from 3 mm to 2 cm across and frequently had five steles (hence the name of the genus and of the group). Each stele had its own cambium, which in young stems was equally active all round; but in older stems the secondary wood was markedly excentric, with a much greater development towards the centre of the stem (Fig. 19B). In addition, there were five much smaller steles (not shown in Fig. 19B) which Sahni suggests might have belonged to short shoots. These alternated with the five main steles. Five was the commonest number of steles, but there were sometimes six, and Vishnu-Mittre was able to show that the number varied along the length of the stem. Thus, Fig. 19C, based on a series of transverse sections, shows a reconstruction of a stelar system with three at the lower end (together with a branch-trace) and six at the top. In *Nipanioxylon Guptai* a single concentric vascular cylinder at one level would give rise to a ring of five to seven separate steles at a higher level and, higher still, these would reunite again to form a single cylinder.

The secondary wood was very compact and very similar to that of a modern *Araucaria*. Thus, the tracheids had uniseriate or bi-seriate bordered pits on the radial walls and the wood-rays were uniseriate, ranging in height from one to five cells. Growth-rings are particularly noticeable but whether these represent annual increments cannot, of course, be discovered.

The leaves, *Nipaniophyllum Raoi*, were borne on short lateral branches (Fig. 19A) and were originally described under the name *Taeniopteris*, because they were strap-shaped, with a well-marked midrib from which the lateral veins arose at an obtuse angle, remaining for the most part unbranched and without anastomoses (occasional dichotomies and very rare anastomoses did, however, occur). Perhaps the most interesting feature of the leaves was the detailed anatomy of the leaf-traces. These had two distinct regions of xylem, one centripetal and the other centrifugal, as in cycads (a condition sometimes described as 'diploxylic'). There were about six of these entering the base of each leaf, each having a separate origin in the stem from one or other of two main steles. The stomata were described by Sahni as being syndetocheilic, like those of the Bennettitales, because of the occurrence of two subsidiary cells alongside the guard-cells. However, Vishnu-Mittre[191] subsequently showed that this conclusion is unjustified, for only a proportion of the stomata were like this. Others had up to six or seven subsidiary cells, and must have been haplocheilic, therefore, like those of cycads.

Although organic connection has not yet been demonstrated, it seems fairly certain that both the seed-bearing and pollen-bearing organs were borne terminally on the short shoots. Fig. 19F shows a reconstruction by Sahni of the seed-bearing structures, *Carnoconites*, which must have looked very much like stalked mulberries. Vishnu-Mittre agrees that this reconstruction is substantially correct, but he maintains that the stalks were unbranched. The mulberry-like 'infructescences' of *C. compactum* were up to 1·8 cm long and contained some twenty sessile ovules attached to a central receptacle. There were neither interseminal scales, nor anything that could be called a sporophyll (the term 'cone' is, therefore, not really appropriate, although Sahni described them as such). Each ovule had a single integument with a hard almond-shaped stony inner region, surrounded by a fleshy sarcotesta, and the integument was free from the nucellus. The infructescences described under the name *C. laxum* were longer (up to 3 cm long) and narrower (only 5 mm across), but, otherwise, were very similar to those of

C. compactum, and probably belonged to *Nipanioxylon Guptai.*

Fig. 19D is based upon a reconstruction by Vishnu-Mittre[191] of the pollen-bearing organs *Sahnia nipaniensis* and shows the way in which the sporangiophores (about twenty-four in number), arranged in a single whorl round a dome-shaped receptacle, were fused at the base into a disc.

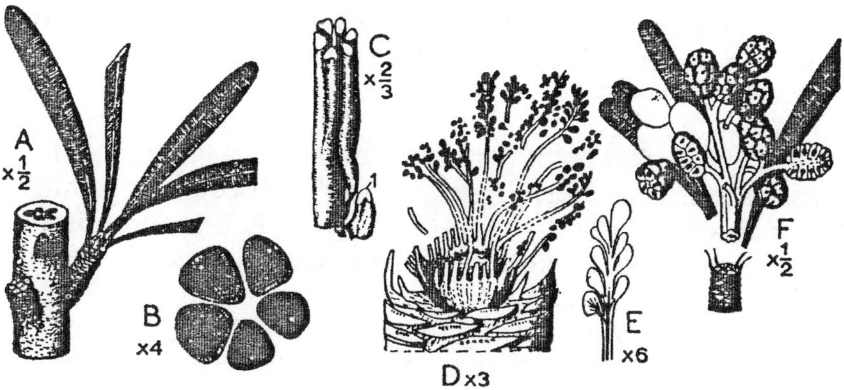

FIG. 19

Pentoxylales

Pentoxylon Sahnii: A, reconstruction of stem and leaves (*Nipaniophyllum Raoi*); B, t.s. stele; C, reconstruction of stele (1 = branch trace). *Carnoconites compactum:* F, reconstruction of female cones. *Sahnia nipaniensis:* D, reconstruction of male 'flower'; E, detached microsporophyll.

(A, B, F, after Sahni; C, E, Vishnu-Mittre; D, based on Vishnu-Mittre)

The microsporangia were unilocular and pear-shaped, and were borne on short branches, arranged spirally on the sporangiophores (Fig. 19E). Each sporangium received a vascular bundle which divided to give several radiating branches. The pollen-grains were boat-shaped and had a single longitudinal furrow.

The correct phylogenetic placing of the Pentoxylales is a most fascinating problem. Their polystelic stems resembled those of some of the Palaeozoic Medullosaceae, yet the secondary wood of *Pentoxylon* was pycnoxylic. Both the male and female organs were stachyosporous (i.e. they were borne on stems, rather than on leaves). The way in which the ovules were borne was not unlike that of the Bennettitales, but there were no interseminal scales. The whorled microsporangiophores were superficially similar to those of the Bennettitales, but their detailed structure was very different. In their leaf-trace anatomy and their stomatal structure the Pentoxylales were like the Cycadales, but few other similarities can be seen. They are, indeed, an enigmatic group.

6

Cycadales

Woody plants with stems unbranched or with occasional adventitious branching. Manoxylic. Mucilage canals in pith and cortex. Some genera with additional co-axial vascular cylinders. Leaves large, pinnate (rarely bi-pinnate). Leaf-trace 'diploxylic' (except in Nilssoniaceae). Dioecious. Reproductive organs in cones (except female *Cycas*). Cones terminal or lateral. Megasporophylls with sterile tips and 8 – 2 orthotropous ovules. Seeds large. Microsporophylls scale-like or peltate with pollen-sacs on the abaxial side. Sperm with spiral band of flagella.

Nilssoniaceae*

Leaves: *Nilssonia**, *Pseudoctenis**
Seed-bearing organs: *Beania**
Pollen-bearing organs: *Androstrobus**

Cycadaceae

*Palaeocycas** (leaves = *Bjuvia**)
Cycas
Microcycas, Dioon, Ceratozamia, Zamia, Stangeria, Bowenia, Encephalartos, Macrozamia

The Cycadales first appeared in Upper Triassic times and are the only group out of all the Cycadopsida to have survived until the present day. Among living plants, therefore, they are extremely ancient and it is perhaps not surprising that they exhibit a number of very primitive features, both in their morphology and in their life cycle.

Palaeocycas integer was described by Florin[61] from Upper Triassic rocks of Sweden, the only parts that are known being the leaves (*Bjuvia simplex*) and some structures (to which the name *P. integer* properly applies) that are believed to have borne seeds. The leaves were large, about 1 m long and 20 cm broad, and had a simple lamina rather like that of a banana, while the seed-bearing structures were much smaller, about 13 cm long and 5 cm broad (Fig. 20B). Florin's reconstruction of the plant is illustrated in Fig. 20A and shows the leaves arranged in a crown at the apex of a stout trunk, with a loose aggregation of seed-bearing structures in the centre. Arnold[36] remarked on the obvious resemblance of the plant to a modern *Cycas*, but this is precisely what Florin intended, for he drew his reconstruction so as to look like *Cycas*. As Harris emphasises, the trunk is completely unknown and, therefore, the manner in which the leaves were borne on it is not known. It is not even certain that the supposed reproductive organs did in fact bear seeds, for all that can be seen are scars where four seeds might have been attached. That *Palaeocycas* and *Bjuvia* belonged to one and the same plant is fairly certain because of the similarity of their cuticular structure (shape of epidermal cells, guard-cells, subsidiary cells, etc.), and that they belonged to a cycad is, likewise, fairly certain. Florin's reconstruction, then, was a reasonable piece of guesswork, but it must be borne in mind that it is no more than this. The same, of course, might be said of many other reconstructions of fossil plants.

Having drawn attention to the dangers inherent in such reconstructions, Harris[99] then illustrates his own reconstruction of *Beania* (Fig. 20C), a genus of fossil cycads from the Jurassic of Yorkshire. The name is properly applied to seed-bearing structures, two species of which are known, in which peltate sporophylls, each bearing two ovules, are arranged spirally in a loose cone around a central axis, although the name has also been used for detached seeds. In *B. gracilis* (Fig. 20D) the cone was up to 10 cm long, with sporophylls ranging from 1·5 to 2·5 cm long. The seeds, when mature, ranged in size from 7 × 7 mm to 16 × 13 mm and, on maceration, yield four cutinised layers, whose distribution indicates that the integument was fused to the nucellus throughout the lower two-thirds of the seed, but free from it at the apex (Fig. 20E). There was an outer fleshy layer, a middle stony layer and an inner fleshy layer, in which there is evidence of vascular bundles running longitudinally up to the level at which the integument became free from the nucellus.

Recent investigations[182] have demonstrated the remains of pollen-grains inside the micropyle which were identical with those obtained from the male cones, known as *Androstrobus manis*. These cones (Fig. 20F) were much more compact than the ovuliferous ones (about 5 cm long and 2 cm broad) and had a large number of spirally arranged peltate sporophylls, on the abaxial side of which were scattered numerous finger-like pollen-sacs (Figs. 20G and H). There can be little doubt that these cones belonged to the same plant as *Beania gracilis*, because of their frequent association in the rocks and because of their stomatal structure, quite apart from the similarity of the pollen-grains, and, for similar reasons, it is believed that the leaf was *Nilssonia compta*. This was a broadly linear leaf, up to 40 cm long and 9 cm wide, whose lamina was cut in an irregular fashion into truncate segments, with parallel veins at right angles to the rachis. In addition, there were some scale-leaves, *Deltolepis*, 'which bridge the structural gap between the other organs'.[99]

Beania Mamayi was similar to *B. gracilis*, but the distal part of the sporophyll was scarcely expanded at all (it would almost be described as a bifid sporangiophore). It is associated in the rocks with *Nilssonia tenuinervis* and with *Androstrobus Wonnacottii*. Harris argues that the cones were so lax that they are much more likely to have been pendulous than erect and, therefore, that they were probably borne on a plant with an aerial stem. Furthermore, the leaf-base of *N. tenuinervis* was so wide that the stem must have been stout, and it is in this way that he arrives at his reconstruction of the female plant (Fig. 20C).

Androstrobus prisma is associated with *Pseudoctenis Lanei*, a leaf that was much more like those of modern cycads, in being pinnate, with narrow lanceolate pinnae.

At the present day the Cycadales are represented by nine genera, comprising some sixty-five species—all that remain

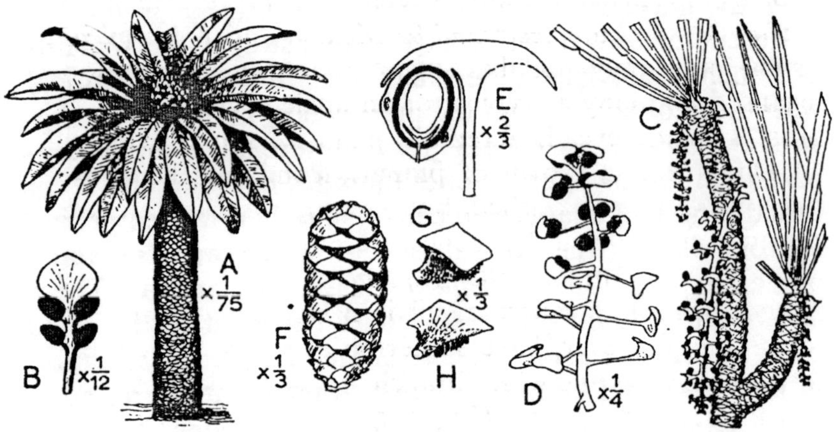

FIG. 20

Fossil cycads

Palaeocycas: A, reconstruction of plant, bearing leaves (*Bjuvia simplex*) and female sporangiophores (*Palaeocycas integer*)—the stem is imaginary; B, single sporangiophore—the seeds are imaginary.

Beania: C, reconstruction of *B. Mamayi*, bearing young and old cones and leaves (*Nilssonia tenuinervis*)—the stem is imaginary; D, female cone of *B. gracilis*; E, female sporangiophore of *B. gracilis* in l.s.

Androstobus: F, reconstruction of male cone (*A. manis*); G. H, male sporophyll, in different views.

(A, B, after Florin; C–H, Harris)

of a group that has been in existence for at least 200,000,000 years. Having been almost world-wide in distribution in past ages, they are now restricted to three regions, Central America, South Africa and eastern Asia with Australia. Within each of these areas there is one genus of relatively wide distribution, while the rest are restricted, some to extremely small areas. Thus, *Cycas* (15 spp.) extends from Japan to Queensland and occurs also in India and Madagascar, while *Bowenia* (1 sp.) and *Macrozamia* (9 spp.) are restricted to Queensland and north-east Australia; *Encephalartos* (14 spp.) is widely distributed in South Africa, while *Stangeria* (1 sp.) is restricted to Natal; *Zamia* (30 spp.) extends from Florida to Chile, whereas *Ceratozamia* (2 spp.) and *Dioon* (3 spp.) are restricted to Mexico and *Microcycas* (1 sp.) is found only in Cuba. Such a pattern of distribution is regarded by many as typical of an ancient group.

Most modern cycads resemble palm-trees in having a stout trunk with a crown of pinnate fronds at the apex. Mostly, they are unbranched, but in some species there are occasional adventitious branches. Several species attain a height of 10 or 15 m, e.g. *Dioon spinulosum* and *Microcycas calocoma*, while the tallest of all is *Macrozamia Hopei* at 18 m. Some, however, have subterranean stems which are short and tuberous and which branch in an irregular manner, e.g. *Stangeria* and *Bowenia* (Fig. 21C) and several species of *Zamia* (Fig. 21B), *Macrozamia* and *Encephalartos*. In the sterile condition some of these bear a striking resemblance to ferns. Indeed, *Stangeria* was classed with the ferns for a long time, for its pinnae are relatively broad and, unlike those of other cycads, have pinnate venation. The likeness to a fern is further enhanced by the sub-circinate vernation of the rachis of the young frond. This also occurs in *Bowenia* and *Ceratozamia*, whereas in *Cycas* the leaflets are circinate. In all the remaining genera, however, the young fronds are quite straight, and so are the leaflets. The only genus, other than *Stangeria*, whose pinnae have a midrib is *Cycas*, but here there are no other veins in the pinna at all. In the remaining genera each pinnae has several parallel, or dichotomous, veins.

The largest fronds are those of *Cycas circinalis*, which may be as much as 3 m long, and the smallest are those of *Zamia pygmaea*, about 5 cm long. They all show strongly xeromorphic structure (even those species which grow in dense forests) in having a heavily cutinised epidermis and deeply sunken stomata. The most interesting anatomical feature from the evolutionary point of view, however, is the structure of the leaf-traces and veins, in which there are two distinct kinds of xylem (Fig. 21G). One is a roughly triangular area of centripetal xylem (1), with a single protoxylem region; the other is separated from it by parenchyma and consists of an arc of tracheids (2) associated with phloem (3) in such a way as to give the appearance of being secondary in origin. Whether this should be regarded as secondary wood or as centrifugal primary xylem is a matter of debate.[120] The important point to notice is the presence of centripetal wood, for this is regarded as a primitive character that has survived from some ancient protostelic ancestor. Centripetal wood has also been demonstrated in the peduncle of the cone in *Stangeria*, *Bowenia* and some species of *Zamia* and *Ceratozamia*,[153] as well as in the sporophylls. Young seedlings also exhibit distinctly mesarch wood in their stem anatomy, even though adult stems are entirely endarch.

Considering the size of many cycad trunks, the amount of secondary wood is surprisingly small, but the persistent leaf-bases make up for this by contributing to the mechanical strength of the trunk. Most species have a single persistent cambium (Fig. 21F), but in *Cycas* and some species of *Macrozamia* and *Encephalartos* a succession of cambia gives rise to co-axial cylinders of secondary xylem and phloem (Fig. 21E). *Zamia* and *Stangeria* are remarkable in that the secondary wood consists of scalariform tracheids—another feature regarded as primitive. Elsewhere, the tracheids have multiseriate bordered pits. Not only is the wood limited in extent, it is also very diffuse, with many rays, one to seven cells wide, and wide leaf-gaps.

The course of the leaf-traces through the cortex is peculiar and gives rise to the 'girdling bundles' that can be seen in transverse sections of the stem (Fig. 21F). The leaf-trace often passes halfway round the stem almost horizontally, so that it enters the leaf almost opposite its point of origin, being joined at intervals by other traces, so that each leaf receives a number of bundles. This is in marked contrast to the Bennettitales, where the leaf-traces pass straight out through the cortex, but it is not unique in the plant kingdom, for a somewhat similar arrangement is found in some flowering plants.

All cycads are dioecious and, in one species at least, the sex of the individual plant is known to be determined by X and Y sex chromosomes.[26] In all genera except *Cycas*, the reproductive organs are borne in compact cones which, in many genera, are borne one at a time at the apex of the stem. Such genera show peculiar features in their internal anatomy, called 'cone-domes'. These can be seen most readily in a longitudinal slice down the centre of the stem and consist of domes of vascular tissue crossing the central pith at intervals (Fig. 21D). The explanation of these lies in the fact that the stem is really a sympodial system. Every time a cone is formed, the vegetative growth of that part of the stem ceases, because its apical meristem is used up. Further growth then continues by the activity of a new meristem that arises near the base of the peduncle. In the process the old cone-scar is pushed aside and its vascular supply remains as a cone-dome across the centre of the stem. It is a remarkable fact that, although the male plant of *Cycas* is sympodial, the female plant is monopodial (Fig. 21A) and its apical meristem retains its identity throughout the life of the individual (accordingly, it does not exhibit cone-domes).

The apical meristem in all cycads is extremely massive, compared with that of other seed-plants. That of *Cycas revoluta* may be as much as 3,000 μ across. Its organisation (Fig. 21H) is of a very primitive kind, found elsewhere only among Ginkgoales. There is a shallow cap of cells, the

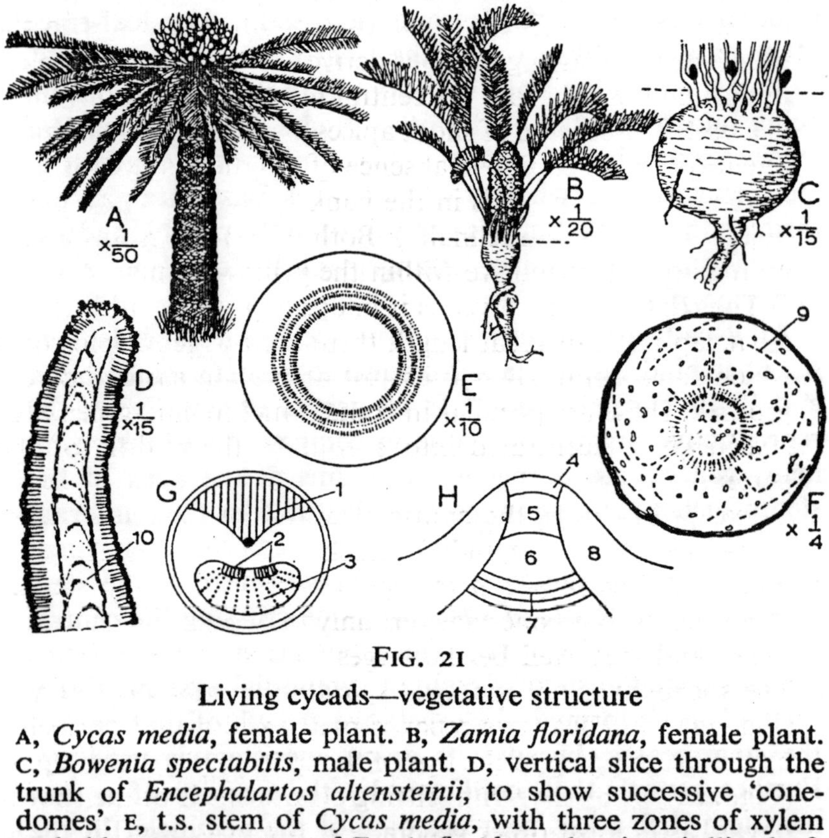

FIG. 21

Living cycads—vegetative structure

A, *Cycas media*, female plant. B, *Zamia floridana*, female plant. C, *Bowenia spectabilis*, male plant. D, vertical slice through the trunk of *Encephalartos altensteinii*, to show successive 'cone-domes'. E, t.s. stem of *Cycas media*, with three zones of xylem and phloem. F, t.s. stem of *Zamia floridana*, showing girdling leaf-traces, scanty zone of secondary wood and abundant mucilage ducts. G, leaf-trace bundle of *Cycas revoluta* (diagrammatic) showing two types of xylem. H, apical meristem of a cycad (diagrammatic).

(1, centripetal xylem; 2, centrifugal xylem or, possibly, secondary wood; 3, regularly arranged phloem; 4, apical initials; 5, sub-apical initials; 6, central mother-cells; 7, pith-rib-meristem; 8, flanking region; 9, girdling bundle; 10, cone-domes)

(A, based on photograph from Chamberlain; B, after Wieland; C, E, F, Chamberlain; G, de Bary; H, based on Johnson)

apical initials (4), which gives rise to a central core and a flanking region. Within the core, three regions can be seen: sub-apical initial cells (5) that are small and densely staining, central mother-cells (6) that are large and lightly staining, and a pith-rib-meristem (7) whose activities give rise to the pith. The flanking region (8) is derived partly from the apical initials and partly from the central mother-cells. Johnson,[104] in a useful discussion of shoot apices of gymnosperms, draws attention to the complete absence of a tunica in cycads, and to the fact that the cells in the flanking region divide periclinally (as well as anticlinally). Both of these features are to be regarded as primitive within the gymnosperms.

The root system of cycads is interesting, for in addition to stout tap-roots of great length there are apogeotropic roots which form coralloid masses just above the surface of the ground. These are peculiar in containing the blue-green alga *Anabaena*, concentrated into a zone in the middle of the cortex.

Except in *Cycas*, the ovules of cycads are borne in compact cones (Figs. 22D and F), which range in size from 2 cm long in *Zamia pygmaea* to 70 cm or more in *Macrozamia Denisonii*, whose cones are certainly the largest in the world today and may well be the largest that have ever existed.[5] The sporophylls are arranged spirally around a central axis (Fig. 22A) and bear two reflexed ovules, which are protected by the overlapping of the tips of the sporophylls, as in *Dioon* (Fig. 22B), or by the tightly fitting peltate heads of the sporophylls, as in most other genera, e.g. *Macrozamia* (Figs. 22C and D) and *Zamia* (Figs. 22E–G).

Throughout the life of the female plant of *Cycas* there is an alternation of fronds and sporophylls, produced by the same apical meristem, and the sporophylls are much more leaf-like than in any other genus (Figs. 22H–K). Furthermore, in some species each sporophyll may bear as many as eight ovules (Fig. 22H), while in others there may be only two (Fig. 22K).

It is widely accepted by morphologists that the *Cycas*

sporophyll is the primitive type among cycads, and that, during the evolution of the group, there has been a reduction in the number of ovules to two, followed by a gradual reduction of the distal region to produce the more advanced peltate types, with *Dioon* representing an intermediate stage. According to this interpretation, female

FIG. 22

Cones of living cycads

Female cones: A, *Dioon edule*, partially dissected; D, *Macrozamia spiralis*; F, *Zamia floridana*.

Female sporophylls: B, *Dioon edule*; C, *Macrozamia spiralis*; E, *Zamia Skinneri*; G, *Z. Fischeri*; H, *Cycas circinalis*; I, *C. revoluta*; J, *C. siamensis*; K, *C. circinalis*, var *papuana*.

Male cone: Q, *Stangeria paradoxa*, cone at the stage of pollen-shedding.

Male sporophylls: L, *Cycas circinalis*; M, *C. Rumphii*; N, *Macrozamia spiralis*; O, *Ceratozamia mexicana*; P, *Zamia furfuracea*.

(B, E, G–L, O, P, after Schuster; C, D, N, Brough and Taylor; F, Wieland)

cycads are phyllosporous. An alternative view, however, is that the peltate sporophyll is the primitive type, and the morphological series is read in the opposite direction as one of gradual proliferation. According to this interpretation they are stachyosporous, and the ovuliferous structures ought to be called sporangiophores, instead of sporophylls. Unfortunately, the fossil record does not resolve this argument, for if *Palaeocycas* has been correctly interpreted, then both types of sporophyll occurred in the Mesozoic, *Beania* representing the type found in most modern cycads.

All the male strobili of cycads (even those of *Cycas*) are compact cones which are either terminal or, sometimes, axillary as in *Macrozamia Moorei*. The microsporophylls are spirally arranged (Fig. 22Q) and bear microsporangia scattered over the lower (abaxial) side. In *Cycas* (Fig. 22L) there may be over a thousand on each sporophyll, but in others the numbers are fewer (Figs. 22M–O), and in *Zamia* (Fig. 22P) there may be as few as five or six. At maturity the sporangia are very similar to those of eusporangiate ferns in having a massive wall, several cells thick, and their origin is typically eusporangiate (Figs. 23J–N). One hypodermal cell (or sometimes several) divides to give a primary wall-cell and a primary sporogenous cell. From these, the thick wall and numerous sporogenous cells are produced, the latter being surrounded by a tapetum which breaks down to produce a plasmodium in which they develop further. Each spore mother-cell undergoes meiosis to give a tetrad of haploid pollen-grains. The haploid number of chromosomes ranges from thirteen in *Microcycas* and eleven in *Cycas* down to eight in *Stangeria* and *Ceratozamia*.

Stages in the development of the ovule of *Bowenia* are illustrated in Figs. 23B–D. Up to the stage shown in Fig. 23B, the nucellus and integument keep pace, but thereafter the integument grows faster, to produce the micropyle. In Fig. 23B meiosis is shown as having been completed to give the single functional megaspore and three abortive megaspores. In Fig. 23C free nuclear divisions are taking place in the periphery of the female prothallus, and this continues until over a thousand nuclei have been formed, before the whole prothallus eventually becomes cellular. Figs. 23F–I show the

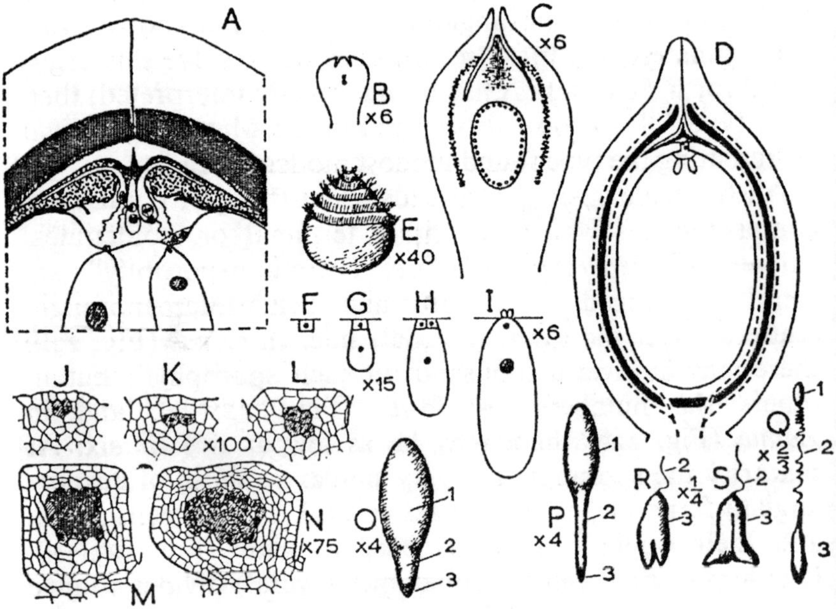

FIG. 23

Stages in the life history of cycads

A, apical region of the ovule of *Dioon edule*, at the time of fertilisation. Five pollen-grains have produced pollen-tubes, two of which have discharged their contents. The left-hand archegonium has been fertilised, but the spermatozoids have not yet entered the right-hand archegonium. B–D, stages in development of the ovule of *Bowenia spectabilis*; B, at the stage of meiosis; C, with free-nuclear divisions in the female prothallus; D, mature (diagrammatic). E, spermatozoid of *Zamia*. F–I, stages in development of the archegonium of *Dioon edule*. J–N, stages in development of the microsporangium of *Stangeria paradoxa* (sporogenous cells shaded). O–S, stages in development of the embryo of *Cycas circinalis*.

(1, egg-sac; 2, suspensor; 3, embryo proper)

(A, F–I, after Chamberlain; B–D, Kershaw; J–N, Lang; O–S, Treub)

stages in development of the archegonia, which consist of four[212] neck-cells and a central cell containing an egg-nucleus and a ventral-canal-nucleus. In most genera there are only three, four or five archegonia, but *Microcycas* is peculiar in having scores, or even hundreds, in each ovule.

It has recently been shown[160] that stomata occur on the apex of the nucellus of *Zamia*. This is a most interesting

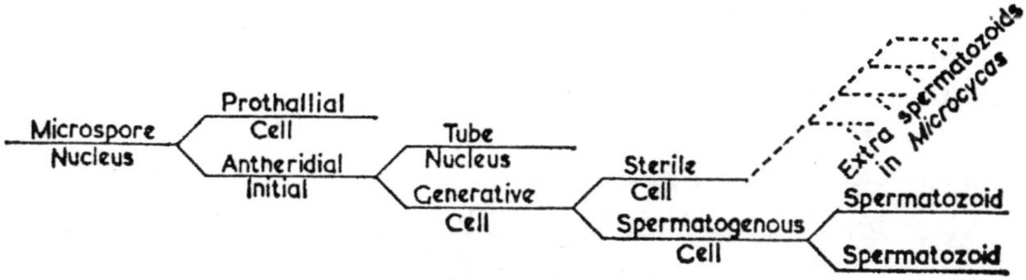

FIG. 24

Stages in the formation of the male prothallus in cycads.

observation, for it is difficult to understand how stomata could perform any useful function in such an enclosed environment. They must surely be a survival from ancient times when the nucellus was exposed, as for example in a seed like *Genomosperma* (Fig. 10G).

Development of the male prothallus commences while the pollen-grains are still inside the microsporangium. The various stages in this process are shown diagrammatically in Fig. 24. The nucleus divides once, so that a single prothallial cell is cut off at one side. The other daughter nucleus, sometimes called the antheridial initial, then divides again to give a tube-nucleus and a generative cell. It is at this stage that the pollen-grain is shed and the ovules are pollinated. The pollen grains are trapped in a drop of fluid secreted by the micropyle and are drawn down into the pollen-chamber. Here a short pollen-tube is produced, which penetrates the sides of the pollen-chamber and which anchors the pollen-grain (Fig. 23A). Meanwhile, the generative cell divides to produce a sterile cell and a spermatogenous cell, which, in turn, divides to give two spermatozoids. As the name suggests, the sterile cell normally plays no further part, but in *Microcycas* it divides repeatedly to produce a succession of ten or eleven spermatogenous cells, each of which then gives rise to two spermatozoids.

The spermatozoids of cycads are the largest known, for they are as much as 300 μ in length. They are pear-shaped and they have a spiral band of flagella (Fig. 23E) which enables them to swim in the drop of fluid that is forcibly extruded into the pollen-chamber when the turgid pollen-tube ruptures. This fluid has such a high osmotic pressure that, when it comes into contact with the neck-cells of the archegonium, they become flaccid, and the spermatozoids then have access to the egg-cell.

After fertilisation has been effected, repeated free nuclear divisions take place, until sixty-four or, more usually, 256, or even over 1,000 nuclei have been formed, before cross-walls are laid down. In some genera, e.g. *Cycas*, *Encephalartos* and *Macrozamia*, the embryo becomes cellular throughout, but in others only the basal region becomes cellular. Stages in the development of the embryo are illustrated in Figs. 23O–S, which show how the chalazal end of the egg-sac becomes elongated as the cells within it divide and elongate to form a suspensor which pushes the 'embryo proper' into the centre of the prothallus. A remarkable feature is the length which the suspensor (2) may attain—sometimes as much as 8 cm when uncoiled. Since there are several archegonia, there are usually several suspensors, which become twisted together into a tangled knot as they grow down into the prothallus; however, only one embryo, as a rule, reaches maturity in any one seed. There are usually two cotyledons (Fig 23S), but occasionally there may be three or only one, and there is a hard coleorhiza, which protects the root-tip until it has penetrated the seed-coat.

When describing the development of the cycad ovule, the term 'integument' has been used here as if it were a single morphological unit, as indeed it appears to be. However, the mature ovule is like that of the Medullosaceae in having two systems of vascular bundles (cf. Figs. 13E and 23D). The outer system consists of several separate unbranched bundles (usually less than a dozen) which run up the outside of the stony layer. The inner system, in the inner fleshy layer, however, forms a complex network of very fine bundles which stops abruptly at the level of separation of the integument and the nucellus. If one were to imagine a phylogenetic fusion of the two integuments of the Medullosaceae, then an ovule very similar to that of the cycads would result. Such a double origin for the cycad integument was suggested as long ago as 1905 by Stopes,[169] but whether the Medullosaceae were their ancestors is still a matter for debate. A connection with the pteridosperms is often claimed, but, one suspects, *faute de mieux* rather than with any firm conviction, and Arnold[36] remarks that 'it is perfectly clear that the cycads are in no sense merely modern pteridosperms' and that 'as yet, we are unable to identify the pteridosperm that served as the ancestor of the cycads'. While *Cycas* is the favourite choice as the most primitive living cycad, because of its female sporophylls, it should be remembered that *Stangeria* shows by far the greatest number of other primitive and fern-like characters, yet in its habit and mode of growth it is very unlike any known pteridosperm.

Haploid chromosome numbers have been recorded as follows: 13 in *Microcycas*; 11 in *Cycas*; 8 in *Stangeria*, *Ceratozamia* and *Zamia* spp.; 9 in the rest.[108] Because of their higher number and 'asymmetrical or specialised karyotypes', Khoshoo[109] regards *Cycas*, and in particular *Microcycas*, as relatively advanced, *Ceratozamia* and *Zamia* being unspecialised and primitive.

The Cycadopsida, described in the previous five chapters, have been seen to be a group whose importance in the vegetation of the world has diminished steadily since Mesozoic times. In contrast, the Coniferopsida, to be described in the next four chapters, by reason of their greater size, have been dominant plants in the world's vegetation from the Carboniferous onwards, and even today they dominate an area of some 4,000,000 square miles, or roughly 8 per cent of the earth's surface. The four groups, Cordaitales, Coniferales, Taxales and Ginkgoales, characterised by their pycnoxylic wood, 'microphyllous' habit and bilaterally symmetrical seeds, are now believed to be distantly related—largely as a result of the work of Florin, whose research over a period of some thirty years represents one of the most outstanding achievements in the history of palaeobotany, solving, for all time it would seem, the vexed question of the nature of the ovuliferous cone-scale of conifers.

7

Cordaitales

Mostly tall trees with slender trunks and a crown of branches. Primary wood scanty. Secondary wood mostly pycnoxylic. Leaves spirally arranged, simple, up to 1 m long, grass-like or paddle-shaped, with parallel venation. Cones compound, unisexual, consisting of a main axis with bracts subtending secondary fertile shoots bearing sterile and fertile appendages. Female fertile appendages with one to four ovules. Male fertile appendages with four to six terminal pollen-sacs. Seeds bilateral.

Eristophytaceae*

Eristophyton, *Endoxylon*, *Bilignea*

Cordaitaceae*

Leaves: *Cordaites*

Stems: *Mesoxylon* (= *Cordaites*?), *Metacordaites*, *Parapitys*, *Caenoxylon*, *Mesopitys*, *Cordaicladus*, *Artisia*

Roots: *Amyelon*

Cones: *Cordaitanthus* (= *Cordaianthus*)

Seeds: *Cardiocarpus* (= *Cordaicarpus*, = *Samaropsis*?), *Mitrospermum*, *Kamaraspermum*

Poroxylaceae*

Stems: *Poroxylon*

Seeds: *Rhabdospermum*

As explained in Chapter 2, the three fossil-stem genera, *Eristophyton*, *Endoxylon* and *Bilignea*, were at one time placed in the Calamopityaceae, but, because of their pycnoxylic wood, Lacey[117] believes that their affinities are with the Cordaitales. All three genera had a small number of circummedullary strands. Thus, *Eristophyton* (Fig. 4C) had about five, which became smaller as they passed downwards from the leaves and which changed from mesarch to endarch before dying away altogether. *Endoxylon* was similar but the strands were endarch throughout. In both these genera the central tissue was pure pith, but in *Bilignea* it consisted of short tracheids. In all three genera the arrangement of the primary strands suggests a phyllotaxy of $\frac{5}{13}$, which in turn suggests that the leaves were not small and crowded; this is why these three genera were regarded as pteridosperms. However, the pycnoxylic secondary wood suggests otherwise. It is, of course, possible that they might have occupied a position intermediate between the Cycadopsida and the Coniferopsida, but, until their leaves and reproductive organs have been discovered, their true affinities must remain in doubt.

The Cordaitaceae reached their greatest development in the Upper Carboniferous, when they formed vast forests of tall stately trees. Their trunks attained a height of at least 30 m, with a crown of branches near the summit (Fig. 25A). Their leaves were spirally arranged and were either grass-like or paddle-shaped, up to 1 m long and 15 cm broad (Fig. 25B). Several sub-genera have been suggested, on the basis of their size and shape, e.g. *Eu-Cordaites*, broad with a blunt tip (Fig. 25A1); *Dory-Cordaites*, broad with a pointed tip (Fig. 25A2); *Poa-Cordaites*, grass-like and less than 3 cm wide (Fig. 25A3). Many species have been described, differing from each other in their internal anatomy, some having well-developed strengthening 'girders' of fibrous cells, but they were all alike in having centripetal xylem in their veins. Most species had centrifugal xylem in addition, and were thus diploxylic like the cycads, and the Pentoxylales.

Each leaf was supplied by two traces—an arrangement which has sometimes been likened to that in *Ginkgo*, but there is a difference in the origin of the two traces. Whereas in *Ginkgo* they arise from two separate protoxylems in the stem, in the Cordaitaceae they arose from a single protoxylem. Then, as they passed out through the cortex, they subdivided so that many traces entered the leaf-base.

The secondary wood of the Cordaitaceae was extremely like that of a modern *Araucaria* in having tracheids with multiseriate bordered pits on the radial walls and wood-rays that were mostly only one cell wide. The primary wood formed a narrow zone, in contact with the secondary wood, with a large number of protoxylems (Fig. 4E). Several genera have been distinguished according to whether centripetal wood is present or not and according to whether the pith is septate or not, but the validity of these distinctions has been questioned.[188] Thus, *Cordaites* and *Mesoxylon* both had septate pith, but the former was stated to be endarch and the latter mesarch. However, Traverse believes that careful examination by modern techniques will show that even in *Cordaites* there was some centripetal wood, however small the amount. *Metacordaites* was defined as having continuous pith and endarch primary wood, whereas *Parapitys* had mesarch (except where the strands were about to die out as they passed downwards in the stem). The two Permian genera *Caenoxylon* and *Mesopitys* had sclerotic regions in the pith, as also did *Eristophyton*, and had endarch primary wood.

The name *Artisia* is given to casts exhibiting septate pith, while roots are known as *Amyelon*. The latter were very similar indeed to those of modern conifers—they were diarch or triarch and apparently contained endotrophic mycorrhizal fungi.[83]

The reproductive organs, both male and female, have been widely known as *Cordaianthus*, but it has recently been shown that the name *Cordaitanthus* takes priority by one year.[73] Their structure is now known in considerable detail as a result of the work of Grand'Eury[79] and, more especially, of Florin.[67] Both the male and the female cones of the Cordaitaceae were built up in the same way and consisted of a main axis bearing distichous bracts (Fig. 25C). These subtended short secondary fertile shoots bearing spirally arranged sterile and fertile appendages which formed part of one and the same phyllotactic spiral.

Fig. 25D illustrates the structure of the female cone of *Cordaitanthus pseudofluitans*, a relatively primitive species, in which the fertile appendages were long and branched, and bore several ovules. In certain views (Fig. 25E) the integument appears to have been formed from the upgrowth round the nucellus of two halves, suggesting a fundamental bilateral symmetry. Fig. 25G is of a more advanced species, *C. Williamsonii*, where each fertile appendage was very short and bore only one ovule which appears to have been buried among the sterile appendages.

Fig. 25H illustrates part of the male cone *Cordaitanthus concinnus* and Fig. 25J a longitudinal section through the secondary fertile shoot of *C. Penjonii*. The fertile appendages (8) as well as the sterile ones (7) were flattened (Fig. 25F), and terminated in six elongated pollen-sacs. However, the dichotomies of the vascular bundle supplying the pollen-sacs suggests that the appendages had evolved from branching telomic structures, even though the term 'sporophyll' is sometimes applied to them.

Several genera of flattened seeds are believed to have belonged to members of the Cordaitaceae because of their frequent association with leaves, etc. All had a very simple pollen-chamber, and the integument was free from the nucellus except at the chalazal end. Figs. 25K and L illustrate the appearance of *Mitrospermum compressum*[32] as if cut through the major and minor planes respectively. In this genus the vascular bundle entering the base of the seed penetrated the stony layer before sending out the two integument bundles. In *Cardiocarpus* the main bundle which entered the base of the seed, after giving off the two integument bundles, continued into the base of the nucellus, where it formed a tracheidal plate from which arose numerous discrete bundles that ran up in the peripheral part of the nucellus.[163] The name *Samaropsis* is used for flattened cordate seeds in which the wing of the sarcotesta was particularly wide and membranous, but whether all such seeds belonged to members of the Cordaitaceae is not certain. Thus, Long[122] suggests that *Samaropsis bicaudata* is merely a compression form of *Lyrasperma scotica* whose affinities are lagenostomalean. *Kamaraspermum*[106] had a peculiar chamber at the chalazal end, which may perhaps have functioned as a flotation device.

Sometimes such seeds are found with a well-preserved female prothallus and sometimes, even, with visible archegonia.[30] It is also common to find pollen-grains in the micropyle or in the pollen-chamber. They are characterised by having an air-sac which almost completely surrounded the grain, between the exine and the intine, and several cells within the body of the grain which, presumably, were male prothallial cells (Fig. 251),[63] like those in the pollen-grains of modern *Araucaria*.

The Poroxylaceae, represented by the single genus *Poroxylon* of Permo-Carboniferous age, was another interesting 'bridging' group, like the Eristophytaceae, exhibiting a combination of pteridosperm-like and cordaitalean characters. Thus, its stem anatomy was very similar to that of *Lyginopteris*. The secondary wood was poorly developed and its wood-rays were as much as three cells wide. In contact with the secondary wood were more than a dozen exarch primary strands, and there was a wide pith in the centre. Even the cortex was like that of a pteridosperm, with its bands of sclerenchyma. In fact, were nothing known of its leaves, this stem genus would certainly be classified with the pteridosperms. However, the leaves of *Poroxylon* were like those of the Cordaitaceae and were up to 1 m long. Furthermore, the seeds associated with it, *Rhabdospermum*,[106] were very similar to *Cardiocarpus*, both in overall shape and

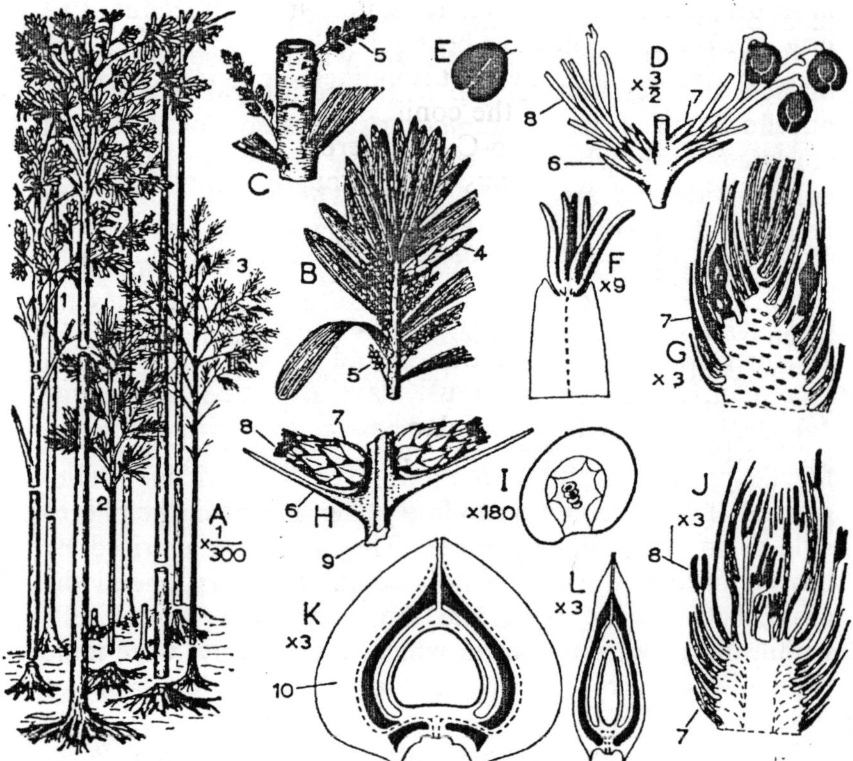

FIG. 25

Cordaitales

A, reconstruction of a forest of *Cordaites* (three sub-genera are included: 1, *Eu-Cordaites*; 2, *Dory-Cordaites*; 3, *Poa-Cordaites*). B and C, fertile branch of *Cordaites laevis*, bearing inflorescences (*Cordaitanthus*). D, part of female inflorescence *Cordaitanthus pseudofluitans*, with two dwarf shoots each in the axil of a bract. E, ovule of *C. pseudofluitans*, seen in a direction at right angles to that in D. F, single male fertile appendage ('microsporophyll') of *C. Penjonii*. G, l.s. female dwarf shoot of *C. Williamsonii*. H, part of male inflorescence of *C. concinnus*, with two dwarf shoots each in the axil of a bract. I, pollen-grain of *Cordaitanthus* sp. showing six prothallial cells and four spermatogenous cells. J, l.s. male dwarf shoot of *C. Penjonii*. K and L, l.s. seed of *Mitrospermum compressum* in two planes at right angles to each other.

in having a number of vascular bundles in the nucellus. Just how *Poroxylon* might have been related to the Cordaitaceae is not clear, but in any event it is unlikely to have played any part in the evolution of the conifers, for these were already in existence by the Permo-Carboniferous.

The presence of numerous vascular bundles in the nucellus of *Cardiocarpus* and *Rhabdospermum* is rather puzzling. By analogy with the seeds of the Medullosae it is tempting to suggest that they represent a number of separate lobes of an inner integument which have become completely fused with the nucellus. However, if this should prove to be correct, then Cordaitalean seeds must have evolved from some ancestral type which was fundamentally radial in symmetry, as Smith[163] has pointed out, and the distinction between the Cycadopsida and the Coniferopsida becomes more blurred than ever.

(4, young vegetative bud; 5, inflorescence; 6, bract; 7, sterile appendages of dwarf shoot; 8, fertile appendages; 9, main axis of inflorescence; 10, wing of seed)

(A–C after Grand'Eury; D–F, I, Florin; G, J, Renault; H, Delevoryas; K, L, Arber)

8

Coniferales

Branching woody plants, often with long and short shoots. Secondary wood pycnoxylic, made up of tracheids with large uniseriate (rarely multiseriate) pits on the radial walls, and small wood-rays. Resin canals in leaves, cortex and (sometimes) in wood. Leaves spirally arranged or opposite, rarely whorled, needle-like or scale-like, rarely broad. Reproductive organs unisexual cones. Female cones fundamentally compound; a main axis with ∞–few bract-scales each subtending, or fused with, one ovuliferous scale (= secondary fertile axis with fertile and sterile appendages) bearing ∞–2 ovules (rarely one). Male cones simple, usually with many scale-like microsporophylls with 2–∞ fused or free pollen-sacs. Embryo with 2–∞ cotyledons.

Lebachiaceae*

Lebachia, Ernestiodendron, Walchia, Walchiostrobus, Carpentieria, Buriadia (?)

Voltziaceae*

Pseudovoltzia, Voltziopsis, Ullmannia

Palissyaceae*

Palissya, Stachyotaxus

Pinaceae

Abies, Pseudotsuga, Picea, Larix, Cedrus, Pinus, Tsuga

Taxodiaceae
Sequoia, Sequoiadendron, Metasequoia, Taxodium, Cryptomeria, Sciadopitys, Athrotaxis

Cupressaceae
Cupressus, Chamaecyparis, Thuja, Juniperus, Callitris, Libocedrus, Papuacedrus

Podocarpaceae
Phyllocladus, Dacrydium, Podocarpus, Saxegothaea, Microcachrys, Microstrobos, Acmopyle

Cephalotaxaceae
Cephalotaxus

Araucariaceae
Agathis, Araucaria

The earliest conifers of which traces have been found, the Lebachiaceae, are from late Carboniferous and early Permian deposits. They were, therefore, in existence before the Cordaitales became extinct. Already they had become segregated into distinct geographical groups separated by the Tethys Sea in the Old World and by its counterpart in the New World. At one time the two genera *Lebachia* and *Ernestiodendron* were included in the single genus *Walchia*, until it was realised that they could be distinguished by their stomata. *Walchia* still has to be retained, however, for specimens in which stomatal details cannot be seen, and it is convenient to speak of all three genera as 'the walchias'. They were restricted to the northern hemisphere, having been found in North America, Europe and Asia; and, in marked contrast with the Cordaitales, they branched in a very regular and characteristic manner. They must, in fact, have looked very much like a modern *Araucaria heterophylla*—the Norfolk Island Pine—for the main branches were in whorls of five or six on the trunk, while the secondary branches were in two rows, arranged in a distichous fashion. The leaves on the ultimate branches were spirally arranged and were needle-like or scale-like with a pointed apex (Fig. 26A), but most of those on the trunk and main branches were of a kind unknown in conifers today, for their tips were bifid.

In *Carpentieria*, another northern genus known only from Central Europe and Morocco, all the leaves forked, suggesting a permanently juvenile condition (Fig. 26I). In the 'southern' genus *Buriadia* (from Brazil and India) the leaves were deltoid, with dichotomous venation. Recently found fertile specimens of *Buriadia*[203] suggest, however, that it was not an early conifer at all, for it had stalked ovules among the leaves.

The walchias were monoecious, i.e. male cones and female cones occurred on one and the same individual. The male cones (Fig. 26F) were borne singly at the tips of ultimate branches, while the female cones (Fig. 26B) were borne either in a similar position or, in some species, at the tips of penultimate branches.

In some respects the female cones were like those of the Cordaitales, in having bracts which subtended secondary fertile short shoots, but, instead of being distichous, the bracts were spirally arranged and had bifid apices. There were further differences in the arrangement of the appendages on the secondary fertile shoots, for, whereas in Cordaitales they were spirally arranged, in the walchias there was a tendency towards dorsiventrality, and in *Lebachia* the sterile appendages were mostly on the abaxial side. This tendency was to become more marked in the Upper Permian and persists in all modern conifers, extending even to the vascular system of the ovuliferous scale. Fig. 26C shows how, in *Lebachia elongatus*, each fertile short shoot bore several sterile appendages and one fertile one terminating in a single erect ovule. *Ernestiodendron filiciforme* (Fig. 26D) was markedly different in having several fertile appendages and no sterile ones. Some species of *Walchiostrobus* were intermediate, with several fertile and several sterile appendages, and some are of interest in having had reflexed ovules (Fig. 26E), another feature that has persisted in many modern conifers. As with the Cordaitales, the ovules give indications that the integument was formed by the fusion of two halves —i.e. was bilateral in its symmetry. Unfortunately, we know very little about their internal structure, for no petrified specimens have been found. There are indications, however, that the integument was free from the nucellus, but whether the latter was vascularised is not known.

The male cones of the Lebachiaceae were radically different, both from those of the Cordaitales and from the female cones, for they were simple strobili with a main axis bearing spirally arranged microsporophylls. This is true also

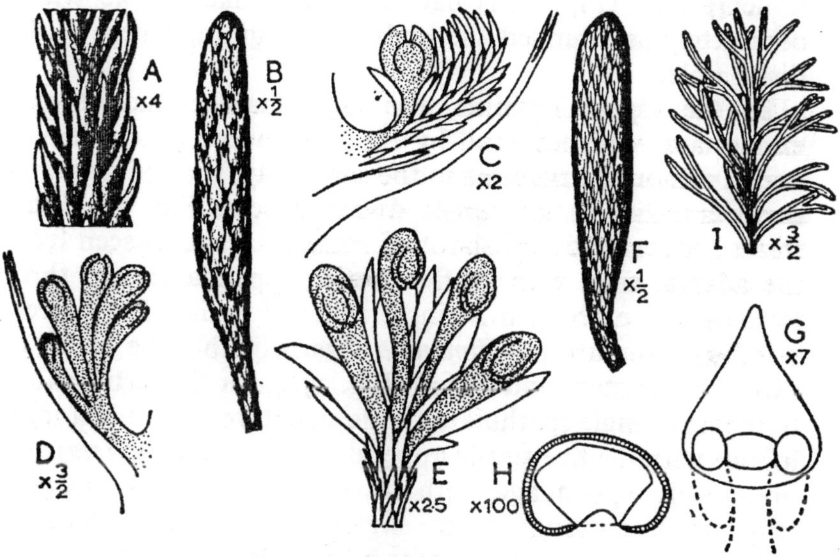

FIG. 26

Lebachiaceae

A, part of leafy axis of *Lebachia piniformis*. B, female cone of *L. piniformis*. C, fertile short shoot and bract-scale of *Walchia* (= *Lebachia*?) *elongatus*. D, fertile short shoot and bract-scale of *Ernestiodendron filiciforme*. E, fertile short shoot of *Walchiostrobus* sp. F, male cone of *L. piniformis*. G, male sporophyll of *L. hypnoides*. H, pollen-grain of *L. piniformis*. I, leafy axis of *Carpentieria frondosa*.

(All after Florin)

of all subsequent conifers, fossil and recent, with the possible exception of some species of *Podocarpus*, but unfortunately no intermediate fossil types are known which might link these simple strobili with the compound cones of the Cordaitales. There is, thus, a very serious gap in our knowledge which must be filled before complete acceptance is possible of Florin's theory concerning the phylogeny of the conifers.[67] In the meantime the assumption is that the complete male cone is homologous, not with the female cone, but with a single fertile dwarf shoot within the female cone. The male sporophylls of the Lebachiaceae had unforked apices and they bore two microsporangia (Fig. 26G). Their pollen-grains (Fig. 26H) had a balloon-like air-sac which completely surrounded the body of the grain, except at the distal pole.

By the Upper Permian, the Lebachiaceae had become extinct and were replaced by the Voltziaceae, whose female cones, although organised in the same way, showed a greater dorsiventrality in the fertile dwarf shoots. Fig. 27A illustrates the fertile dwarf shoot of *Pseudovoltzia*, as seen from the adaxial side, with its five sterile appendages and three fertile ones, each with a single reflexed ovule. The corresponding structure in *Ullmannia* was probably the result of extreme reduction and fusion (Fig. 27B), for the fertile shoot bore just a single spathulate sterile structure (possibly homologous with the five sterile appendages of *Pseudovoltzia*) and one fertile appendage with a reflexed ovule. *Voltziopsis* (Fig. 27C), although from slightly more recent deposits (Triassic), was, nevertheless, at a lower stage of evolution in having five fertile appendages as well as five sterile ones. Moreover, its ovules were erect.

The arrangement of the fertile dwarf shoots of the Voltziaceae, with their sterile and fertile appendages, is clearly comparable with that in Lebachiaceae, and while the Voltziaceae were not necessarily the ancestors of modern conifers, they indicate how all except the Cephalotaxaceae may have originated. The Mesozoic Palissyaceae, on the other hand, were probably more akin to *Ernestiodendron*, lacking sterile appendages. Thus, *Stachyotaxus* (Figs. 27D–F), from the late Triassic of Europe and Greenland, had two ovules on the fertile dwarf shoot which was fused to the subtending bract, while *Palissya* (Figs. 27G and H), of Jurassic age, had about ten ovules on each fertile shoot. Florin suggests that, among modern conifers, the arrangement found in the Cephalotaxaceae is comparable with that in *Stachyotaxus*,

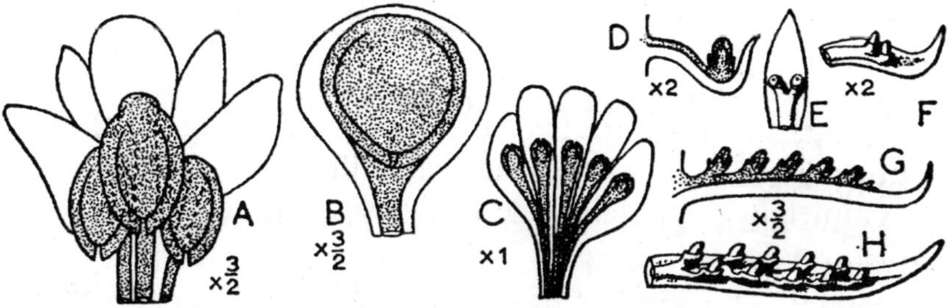

FIG. 27

Voltziaceae and Palissyaceae

Voltziaceae. Adaxial view of fertile short shoots from cones of: A, *Pseudovoltzia Liebiana*; B, *Ullmannia Bronnii*; C, *Voltziopsis africana*.

Palissyaceae. Cone-scale complexes of: *Stachyotaxus elegans* (D, l.s.; E, adaxial view; F, side view) and *Palissya sphenolepis* (G, l.s.; H, side view).

(All after Florin)

for, although there is no sign of fusion between the fertile shoot and the subtending bract in this family, the ovules are borne in pairs on fertile shoots which are completely without a sterile portion. The affinities of this family of modern conifers, therefore, lie with *Ernestiodendron* rather than with *Lebachia*.

Modern conifers

Modern conifers number about 566 species, belonging to some fifty-two genera, and are usually classified into six families (Pinaceae, Taxodiaceae, Cupressaceae, Podocarpaceae, Cephalotaceae and Araucariaceae) on the basis of such morphological characters as: arrangement of parts (leaves, bracts, ovuliferous scales, microsporophylls), spiral or opposite; degree of fusion of ovuliferous scale and subtending bract to form a cone-scale complex; number of pollen-sacs on each microsporophyll; number of ovules on each ovuliferous scale. Not only do coniferous forests cover vast areas of the earth's surface, but, even in mixed forests, conifers are often the tallest trees and project well above the general canopy. The tallest trees in the world, in fact, are conifers, one specimen of *Sequoia sempervirens* being over 112 m high, and the oldest living trees are specimens of *Pinus aristata* over 4,000 years old growing in the mountains of California.[44]

Many of the modern species and genera have a markedly disjunct geographical distribution and there are many endemics. This kind of distribution pattern is usually taken to indicate great antiquity and it is, therefore, not surprising that each of the six families of modern conifers has a fossil record extending back to the Mesozoic. Accordingly, morphologists have searched for evidence of primitiveness, such as the presence of centripetal wood, but, apart from the cotyledons of *Cephalotaxus*, the cladodes of *Phyllocladus*, the male cones of *Saxegothaea* and (perhaps) the ovuliferous scales of *Araucaria*, the conifers are found to be uniformly endarch in their stem and leaf anatomy. Indeed, compared with most other gymnosperms, the conifers are relatively advanced, and it is significant that the group is universally siphonogamous, i.e. fertilisation is by non-motile sperms conveyed to the egg by a pollen-tube.

It is not certain when the Araucariaceae first appeared, for, although the first clearly identifiable remains are from the Triassic, fossil wood of the araucarian type goes back to the Palaeozoic. This type of wood is therefore regarded as the most primitive among living conifers. The tracheids have multiseriate pitting, with the pits often so crowded as to be polygonal in outline. The wood-rays are homogeneous, without resin ducts, and are made up of thin-walled cells with no pits, either on the radial or the tangential walls. The first fossil wood with heterogeneous rays (i.e. consisting of ray parenchyma cells and ray tracheids) did not appear until the Cretaceous, and this feature is therefore regarded as relatively advanced. The most advanced wood of all, according to Greguss,[11] is that of *Pinus* with heterogeneous rays containing resin ducts lined with thin-walled cells, the ray-cells being pitted.

In some other respects, however, the Pinaceae are relatively primitive among conifers. Thus, the apical meristem of *Pinus* shows no clear distinction between tunica and corpus,[104] whereas that of *Araucaria* has a well-marked tuniça, two cells thick in some species, and furthermore has clearly stratified flanking regions. This is because periclinal cell-divisions are absent, not only from the tunica, but also from the outer layers of the corpus. Central mother-cells, however, are absent from *Pinus* and from most of the other conifers that have been examined, except *Pseudotsuga*, *Sequoia* and *Araucaria*. Most of the Pinaceae are like *Pinus* itself in lacking a tunica, and this is true also of the Taxodiaceae, the Cupressaceae and the Podocarpaceae, yet there is stratification of the flanking layer in these families, as well as in some other members of the Pinaceae besides *Pinus*. Much critical work on gymnospermous apices is urgently needed, however, before their full phylogenetic significance can be appreciated. In particular, what is needed is more information about the changes that take place during a single growing season.

The phloem of conifers, like that of other gymnosperms, is without companion-cells. The sieve-cells have sieve-areas confined to the radial walls, and there are also large parenchyma cells with abundant starch and with large nuclei. In *Pinus* there are two kinds of phloem-ray, one dying with the rest of the phloem, the other starch-filled and surviving long after the death of the phloem.[13]

The roots of conifers are commonly diarch, although even pentarch roots occur in some species. Ectotrophic mycorrhizal associations with fungi occur throughout, except in the Araucariaceae, where the fungus is endotrophic, and the Podocarpaceae, which are peculiar in possessing root-nodules rather like those of the Leguminosae. *Taxodium distichum*, the Bald Cypress, when growing in swampy habitats, develops hollow vertical pneumatophores which may be as much as 1 m high and 30 cm across.

On the whole, the leaves of conifers are small and in many genera they are reduced to minute scales, often crowded together so as to look like the scales on a reptile's skin. In such genera, however, the young plant invariably has awl-shaped leaves, and similar 'juvenile leaves' may appear sporadically even on much older parts of the plant. By contrast, the leaves of *Podocarpus* and of *Agathis* may be as long as 30 cm and 17 cm respectively, and as wide as 5 cm. Most conifers are evergreen and in *Araucaria* the leaves may persist and remain green for as long as fifteen years, but *Larix*, *Pseudolarix*, *Taxodium* and *Metasequoia* are peculiar in being deciduous, shedding their leaves each autumn.

In a book of this limited size it would clearly be impossible to describe in detail every one of the fifty-two genera of conifers. There has to be some selection of those which are of special interest. For fuller information the reader is referred to reviews by Florin[70] for geographical distribution, past and present, Florin[67] for cone-scale interpretation, Sterling[164] for male gametophytes, Chowdhury[49] and Doyle[56] for embryology, Khoshoo[108] for chromosome numbers and Doyle[55] for pollination mechanisms. An extensive bibliography of conifer literature was recently prepared by de Ferré.[60]

As might be expected, no one family is primitive in all respects; all show varying combinations of primitive and advanced characters. Accordingly, the order in which they might be described is a matter of arbitrary choice. In this book the Pinaceae are described first because its members exhibit the least degree of fusion between the ovuliferous scale and the subtending bract-scale.

Pinaceae

This is essentially a northern family and consists of ten genera and over 200 species. Not only is it the largest family of conifers, but also the most important economically, for it provides most of the soft-wood timber which constitutes seven-eighths of the world's total lumber. The Pinaceae are characterised as follows: spirally arranged parts; ovuliferous scale free from the bract-scale; ovules two per scale; pollen-sacs two; pollen mostly with two air-bladders.

The morphology and life history of *Pinus* is familiar to most students of botany since the writers of most elementary textbooks choose it as the type to represent the gymnosperms. About ninety species are known. In all of them there are branches of two types, long shoots and short shoots, the foliage leaves being confined to the latter, except in young seedlings. According to species, there are two, three or five needle-shaped leaves in a cluster at the apex of each short shoot (Fig. 28C), but there are also scale-leaves which occur on the long shoots, and it is in their axils that both long and short shoots arise.

Female cones take the place of long shoots and take two years to complete their development. On the main axis of the cone are minute spirally arranged bract-scales, each subtending an ovuliferous scale which bears two reflexed ovules (Figs. 28D and E). The bract-scale grows much more slowly than the ovuliferous scale and in the mature cone is relatively minute. Fig. 28F illustrates the appearance of the cone-scale complex at the stage of pollination (early spring). The ovule has a single integument, fused to the nucellus except at the apex, and there is a single linear tetrad of megaspores, of which only the lowermost is functional. Within the megaspore, free-nuclear divisions occur until some 2,000 nuclei have been formed, but it is not until a year later that the prothallus becomes cellular and archegonia are formed (Fig. 28G). Usually there are three archegonia, but occasionally only one or as many as five, each consisting of an

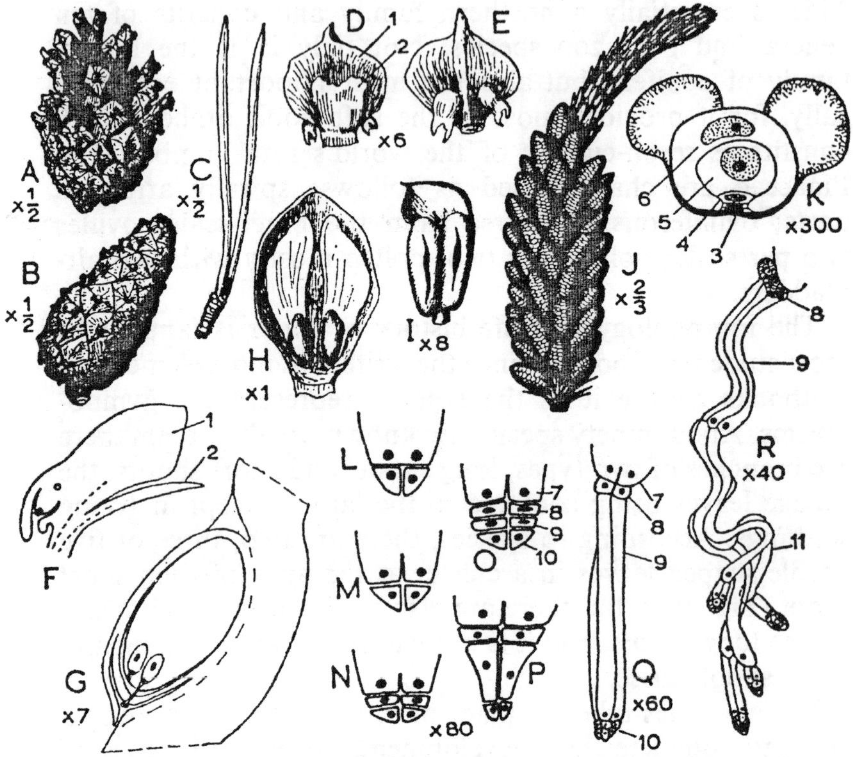

FIG. 28

Pinaceae

Pinus sylvestris: A, female cone after shedding seeds, and B, before; C, short shoot with two needle-leaves; D–H, ovuliferous cone-scale complex (D, abaxial; E, adaxial; F, l.s. at pollination; G, l.s. at fertilisation; H, adaxial view, with two mature seeds). *Pinus nigra*: I, male sporophyll in abaxial view. *Pinus contora:* J, group of male cones. *Abies balsamea:* K, pollen-grain, showing air-bladders and five nuclei. Embryogenesis: L–R, stages in embryo-development of *Pinus sylvestris.*

(1, ovuliferous scale; 2, bract-scale; 3, two prothallial cells; 4, sterile cell; 5, spermatogenous cell; 6, tube-nucleus; 7, U tier; 8, R tier; 9, S tier; 10, E tier; 11, secondary suspensor)

(A, B after Willkomm; D, E, H, Melchior and Werdermann; F, L–P, Coulter and Chamberlain; G, Priestley and Scott; I–K, Chamberlain; Q, McLean and Cook; R, Buchholz)

egg-cell, a ventral canal-cell and a short neck composed of two tiers of four cells.

Pinus is monoecious, and the male cones take the place of short shoots, being crowded together (Fig. 28J) in a manner which invites comparison between the whole group and a single female cone. Each cone bears spirally arranged microsporophylls (Fig. 28I) whose pollen-sacs dehisce by means of a longitudinal slit. At the time the pollen-grain is shed its nucleus has already undergone three divisions, giving two prothallial cells, a tube-nucleus and a generative cell (Fig. 30A). There are two large air-sacs which probably assist the pollen-grain to float to the nucellus, for there is a drop mechanism and the micropyle is inclined downwards. A further division, of the generative cell, gives a sterile cell and a spermatogenous cell, the latter then giving two non-motile sperms. However, it is not until a year after pollination that the pollen-tube actually reaches the archegonium and sheds the male cells into the eggs. Fig. 28K shows the appearance of the pollen-grain at the five-celled stage, the spermatogenous cell apparently standing on a stalk. Indeed, until very recently the sterile cell was termed the 'stalk-cell' and the spermatogenous cell the 'body-cell', the latter being thought of as the body of a stalked antheridium. The successive divisions of the male prothallus in various conifers are illustrated in Fig. 30, where the terminology is that suggested by Sterling.[164]

Stages in the embryology of *Pinus* are illustrated in Figs. 28L–R. The zygote divides twice to give four nuclei which migrate to the chalazal end of the archegonium before dividing again to form two tiers of four cells (M). Both tiers divide once more to give four tiers (O), described from below upwards as: the embryonal (e); suspensor (s); rosette (r); upper (u). This arrangement is peculiar to the Pinaceae, for the rosette tier behaves as a suspensor in other conifers. The cells of the 'u' tier are peculiar in being without an upper wall and are sometimes called open cells. Their nuclei soon degenerate, while the cells of the 's' tier elongate to form a suspensor which pushes the 'e' tier down into the female prothallus (P and Q). The cells of the 'e' tier divide further into proximal secondary suspensors, which split apart, and four distal embryos, which are carried further into the prothallus (R). This splitting process results in what is called 'cleavage polyembryony'. There is also, of course, simple polyembryony, resulting from the fertilisation of several archegonia, but there is a third type, known as 'rosette polyembryony' in some species of *Pinus*. This is brought about by the development of extra embryos from the rosette cells. However, only one embryo normally reaches maturity in any one seed.

At maturity, the female cone is brown and the ovuliferous scales are woody (Fig. 28B). However, some slight amount of growth is still possible, and this causes the scales to spring apart (Fig. 28A), so releasing the ripe seeds. These are winged, the wings being formed from the adaxial surface of the ovuliferous scale (Fig. 28H), and not therefore strictly part of the seed at all.

There are some ten species of *Larix* (Larches) and four of *Cedrus* (Cedars), all of which are like *Pinus* in having long and short shoots. The leaves are not, however, confined to the short shoots, but are scattered along the long shoots also. In *Larix* (Fig. 29K) the short shoots bear terminal tufts of leaves that are renewed each year, while in *Cedrus* there are successive false whorls of leaves on them. In the young female cones of *Larix* the bract is larger than the ovuliferous scale, but in the mature cone it is hidden (Fig. 29L). By contrast, the bracts of *Cedrus* are minute throughout.

In *Abies* (Silver Firs—46 spp.), *Picea* (Spruces—40 spp.), *Tsuga* (Hemlocks—10 spp.) and *Pseudotsuga* (Douglas Firs —7 spp.) there is no differentiation into long and short shoots. *Abies* is immediately recognisable by the disc-like scars which are left when the leaves are shed. *Picea*, on the other hand, has peg-like projections which are left by the falling leaves, and its cones are characteristic in being pendulous at maturity (Fig. 29A) and in persisting for some time after the seeds have been shed. The cones of *Abies* lose their scales, but the cone-axis remains attached. The bracts of *Picea* are minute, but those of *Abies* may even protrude beyond the ovuliferous scales. In *Pseudotsuga* the bracts are much larger, throughout, than the ovuliferous scales and have trifid apices (Fig. 29D–F).

Although the cones of *Picea* are pendulous at maturity, they are erect at the time of pollination, so that the pollen-grains float upwards in the micropylar fluid towards the nucellus, as in *Pinus*. However, in other genera of the Pinaceae no fluid is secreted, and in *Cedrus*, *Tsuga* and *Pseudotsuga* the cone is not geotropically orientated. With the loss of a drop-mechanism, the air-bladders tend to be lost too,[55] and in some genera there is a stigma-like development of the micropyle. In *Larix*, where there are no air-bladders, there is a cushion-like stigmatic outgrowth from one side of the micropyle (Fig. 29M). However, the pollen ultimately finds its way into the micropyle by some process not fully understood. In some species of *Pseudotsuga* the pollen is capable of germinating even when not in contact with the nucellus, and the same is true of *Cedrus*, where there is a one-sided flange on which the pollen alights. Some species of *Tsuga* are remarkable in having pollen-grains that can germinate even on the adaxial face of the ovuliferous scale, the pollen-tube growing along the surface until it can enter the micropyle. The only other conifers which show a similar behaviour belong to the genus *Araucaria*.

Pinus is unique among the Pinaceae in the time taken for its pollen to germinate. In *Abies* only four to five weeks elapse between pollination and fertilisation. In many genera the generative cell has already divided before pollination, with the result that there are five cells within the pollen-grain when it is shed. Another feature of interest is that in most genera the two male gametes are markedly different in size. In such genera the archegonia are widely separated and any one pollen-tube is capable of entering only one archegonium. Where the two male gametes are the same size, the archegonia are usually close together, and two may be fertilised by one and the same pollen-tube.

At one time the early stages of embryogenesis in *Pseudotsuga* were thought to be unique among conifers. It was believed that after the four-nucleate stage only two nuclei moved towards the base, the other two moving to the micropylar end. However, improved techniques of investigation[26a] showed that in general the early development of the proembryo is like that in *Pinus*.

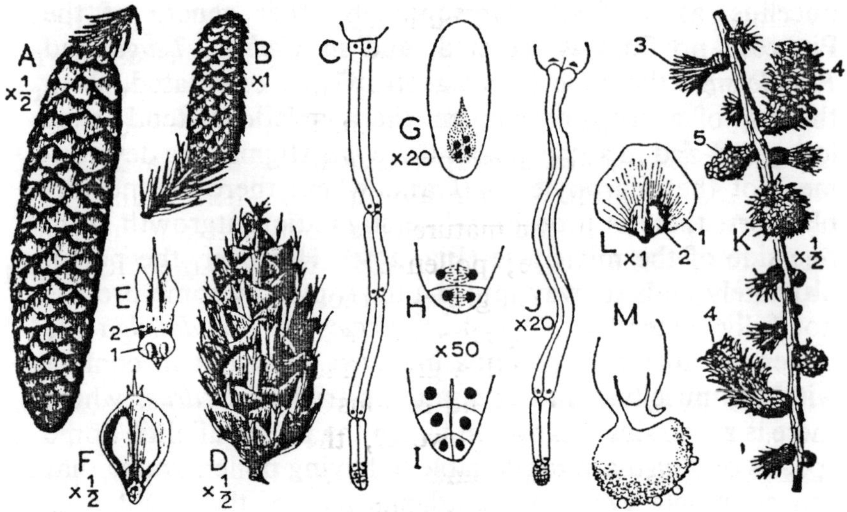

FIG. 29

Pinaceae (continued)

Picea abies: A, mature female cone; B, male cone; C, embryogenesis.

Pseudotsuga Menziesii: D, female cone; E, F, young and old cone-scale complex, in adaxial view; G–J, stages in embryogenesis. *Larix decidua:* K, shoot with young leaves, male and female cones; L, cone-scale complex; M, young ovule (diagrammatic showing stigma-like development of micropyle, with wingless pollen-grains adhering).

(1, ovuliferous scale; 2, bract-scale; 3, young leaves; 4, Female cone; 5, male cone)

(A, B, K, L, after Willkomm; C, J, Buchholz; D–F, Eichler; G–I, Allen; M, Doyle)

Cleavage polyembryony occurs in *Cedrus* and *Tsuga*, as well as in *Pinus*, and also occurs occasionally in *Abies*. In *Picea*, *Larix* and *Pseudotsuga* the four 'e' cells divide several times to form a multicellular mass which then differentiates into a single embryo (Figs. 29C and J).

The basic haploid chromosome number in the Pinaceae is 12, except for *Pseudotsuga*, where it is 13.

Taxodiaceae

This family is of great historical interest, for, whereas today it shows only a relict distribution, at one time its members played an important role in the forests of the northern hemisphere. Of the ten genera, seven are monotypic, and the total number of species is only fifteen. They are characterised by the following features: parts mostly spirally arranged; bract and ovuliferous scale almost free when young, but extensively fused in the mature cone; ovules two to nine on each ovuliferous scale; pollen-sacs two to nine on each microsporophyll; pollen-grains without air-bladders.

Two of the three species of *Taxodium* are limited to southern U.S.A., where they grow in lowland swamps, while the third is found in upland regions from Texas to Guatemala. In earlier times, however, the genus occurred in Canada, Alaska, Europe and western Asia. There are two types of branchlet: one persistent, with axillary buds, near the end of the year's growth, the other without axillary buds and shed along with the leaves, either irregularly or, in *T. distichum*, annually.

Metasequoia, which also sheds its branchlets along with the leaves each autumn, was unknown except as a fossil from Pliocene deposits until in 1948 living specimens of *M. glyptostroboides* (Dawn Redwood) were discovered in a very restricted area of central China. More recent investigation, however, shows that the genus was once widely distributed in North America, Greenland and several Arctic and sub-Arctic islands, as well as in Asia, and that it extended back to the Cretaceous.

Sequoia, represented today by the one species *S. sempervirens* (Californian Redwood), restricted to a narrow coastal belt in the Californian region, was once widely distributed throughout the northern hemisphere. Similarly, *Sequoiadendron* is represented by the one species *S. giganteum* (Big Tree, or Wellingtonia) and is restricted to central California, whereas in past ages it grew also in Greenland and Europe. Its leaves are small and lanceolate, whereas *Sequoia* has two types of leaf—those on the lateral branchlets are linear and

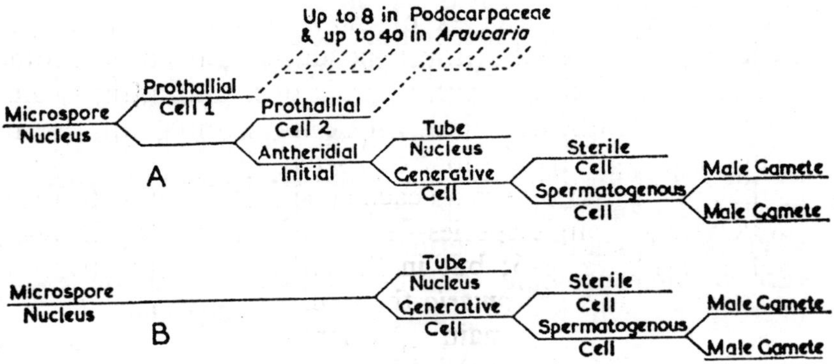

FIG. 30

Stages in development of the male prothallus in conifers

A: Pinaceae, Podocarpaceae and Araucariaceae. B: Cephalotaxaceae, Taxodiaceae and most Cupressaceae.

about 2 cm long, while those on the leading shoots are short and lanceolate like those of *Sequoiadendron*.

Cryptomeria japonica (Japanese Cedar), the only species of the genus, is restricted to Japan and parts of China. Its branches are in whorls, and the branchlets are eventually deciduous, along with the awl-shaped leaves.

Sciadopitys is represented by the one remarkable species *S. verticillata* (Japanese Umbrella Pine), which, although restricted to Japan today, was at one time widespread in Europe too. Its photosynthetic organs, borne in pseudowhorls of ten to thirty, resemble two pine-needles, about 12 cm long, fused face to face. They are sometimes described as cladodes, and certainly they exhibit some stem-like features, for they are subtended by scale-leaves and they sometimes give rise to shoots directly.

Athrotaxis (three species) is the only genus in this family with a southern distribution, being confined to Tasmania. The leaves are small, crowded and overlapping, almost scale-like in one species, and the ultimate branchlets are deciduous as in *Cryptomeria.*

The ovuliferous scale of *Cryptomeria* (Figs. 31H and I) is of particular interest because of the several tooth-like projections (besides the apex of the bract-scale) which may represent the tips of sterile appendages that have become incorporated with the several fertile ones during the evolution of the single ovuliferous scale from something resembling the short fertile shoot of *Pseudovoltzia.*

The number of ovules on each cone-scale is liable to vary, even within a single species—e.g. *Sciadopitys*, 5–9; *Sequoia*, 5–7; *Athrotaxis*, 3–6; but in *Taxodium* there are two. In *Taxodium* and *Cryptomeria* the ovules are erect (Figs. 31G and I), but in the remaining genera described above they are inverted (Fig. 31B). A drop-mechanism is present throughout the family, although the pollen-grains are without air-bladders and the micropyles are not necessarily vertical. Presumably, the drop is actively re-absorbed as soon as any pollen-grains of the right species become trapped in it, thereby drawing them into the micropyle, but little is known of the details. No male prothallial cells are formed at all (Fig. 30B), the pollen-grain nucleus functioning directly as an antheridial initial.

Apart from *Sequoia* and *Sciadopitys*, embryogenesis is fairly uniform throughout the family. Early divisions give rise to three tiers of four cells—an upper 'u' tier, an 's' tier (instead of a rosette tier) and an 'e' tier. Secondary suspensors are formed from the 'e' tier, and cleavage polyembryony occurs throughout. Embryogenesis in *Sequoia* follows a different pattern. Cross-walls are formed, right from the start, to give a four-celled pro-embryo. Then each cell divides into one 's' cell and one 'e' cell (Fig. 31E). The 's' cells elongate and the 'e' cells each give rise to secondary suspensors and an embryo. This pattern of behaviour may be regarded as cleavage polyembryony of a particularly precocious kind.

Sciadopitys is also peculiar in its embryogenesis, for free nuclear divisions proceed until thirty-two, or even sixty-four, nuclei have been formed. These then organise themselves into tiers, as follows: a 'u' tier of 20–30 free nuclei, an 's' tier of 7–13 cells, and an 'e' tier of 12–20 cells. There may also be some secondary suspensor cells, and cleavage polyembryony of a high order occurs, giving 12–28 embryos.

Not only does *Sciadopitys* differ from the other members of the Taxodiaceae in its leaf morphology and in its embryogenesis, but it also differs in its basic chromosome number, for, whereas the basic number for the rest of the family is 11, in *Sciadopitys* it is 10.

Cupressaceae

There are some twenty-two genera, belonging to the Cupressaceae, of which eleven are monotypic, the total number of species being about 148. It is an interesting family in that half the genera are northern and half southern in distribution. All have their parts arranged in an opposite and decussate manner, or in whorls, and there is pronounced fusion between the ovuliferous scale and the bract-scale. The ovules are invariably erect (Fig. 31O) and they vary in number from three to twenty (rarely one or two) per scale. Likewise, there is variation in the number of pollen-sacs (from three to six, rarely two). In some genera the cone-scales overlap (e.g. *Thuja*), but in others they are valvate (e.g. *Cupressus*—Fig. 31N), and they may even be fleshy (e.g. *Juniperus*—Figs. 31J and K), but, in general, it may be said that their female cones are built up in the same basic way as those of the Taxodiaceae.

The pollen-grains are without air-bladders and there is a well-developed drop-mechanism; the glistening drops of fluid can readily be seen with the naked eye because of the erect position of the ovules. The nucleus of the pollen-grain functions as a generative nucleus, there being no male prothallial cells, and in many genera the first division is delayed until after pollination. In some species of *Cupressus* (e.g. *C. sempervirens*) the spermatogenous cell divides rapidly and repeatedly to produce up to twenty male gametes, but, elsewhere, only two male gametes are formed (Fig. 30B). These are equal in size in some genera, but unequal in others and, associated with this difference, a single pollen-tube may branch and enter two archegonia or remain unbranched and enter only one.

Among the northern genera are *Cupressus* (Cypress—20 spp.), with scale-like leaves, and *Chamaecyparis* (6 spp.), also with scale-like leaves but with flattened branch systems in addition. Varieties of the latter with persistent juvenile leaves have sometimes been placed in a separate genus *Retinospora*, but there seems to be little justification for this. *Thuja* (Arbor Vitae—5 spp.) is also northern and, like the preceding genera, has a markedly discontinuous distribution pattern. *Juniperus* (Juniper—70 spp.), however, occupies a continuous broad belt round the northern hemisphere. This may, perhaps, be explained by the fact that its female cones are berry-like and are distributed by birds. In some species the cones are extremely reduced and in the *Microbiota* section there may be just a single cone-scale with a single ovule. In the *Oxycedrus* section of the genus the ovules are displaced laterally, so that they appear to alternate with the cone-scales (Fig. 31L). However, in the *Sabina* section the ovules are placed quite normally in relation to the cone-scales. The leaves in some species are scale-like, although there are always juvenile leaves somewhere on the plant, while in other species all the leaves are of the juvenile type (Fig. 31J).

Among the southern genera are *Callitris* (Cypress-Pines—15 spp.), confined to Australia, Tasmania and New Caledonia, and *Libocedrus* (5 spp.), confined to New Zealand

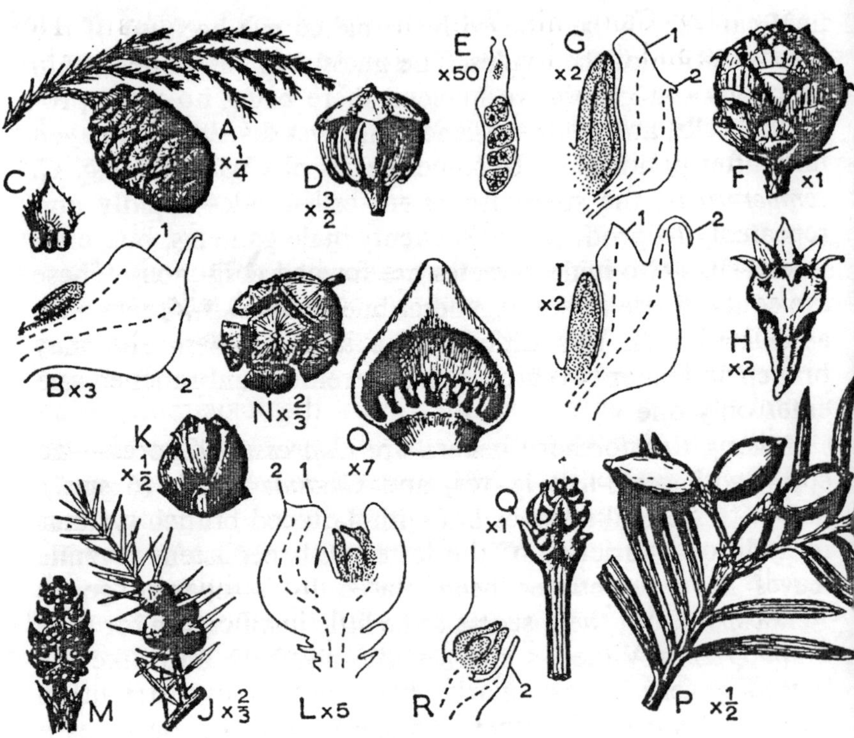

FIG. 31

Taxodiaceae, Cupressaceae and Cephalotaxaceae

Taxodiaceae. *Sequoiadendron giganteum*: A, shoot with female cone; B, l.s. cone-scale complex. *Sequoia sempervirens:* C, male sporophyll; D, cone-scale with ovules; E, embryogenesis. *Taxodium distichum:* F, female cone; G, l.s. cone-scale complex. *Cryptomeria japonica:* H, abaxial side of female cone-scale complex; I, l.s. cone-scale complex.

Cupressaceae. *Juniperus:* J, female shoot of *J. communis*; K, female cone of *J. drupacea*; L, l.s. female cone of *J. communis*; M, male cone of *J. communis*. *Cupressus*: N, female cone of *C. sempervirens*; O, cone-scale of *C. lusitanica*, in adaxial view, showing numerous erect ovules.

Cephalotaxaceae. *Cephalotaxus harringtonia:* P, mature female shoot; Q, young female shoot; R, l.s. ovule and bract-scale.

(1, ovuliferous scale; 2, bract-scale)

(A, S, D, F, K, N, P, after Eichler; B, G, I, Hirmer; E, Buchholz H, Goebel; J, M, Berg and Schmidt; L, Aase; O. Zimmermann; Q, Pilger; R, Sinnott)

and New Caledonia. *Papuacedrus* (3 spp.) occurs in New Guinea and does, in fact, just extend across the equator into the northern hemisphere.

Embryogenesis is fairly uniform in the Cupressaceae. The zygote divides three times, giving rise to eight free nuclei. Wall-formation then occurs and a further nuclear division gives three tiers. However, in *Cupressus sempervirens*, wall-formation takes place at the four-nucleate stage. The middle tier in a three-tiered embryo represents the suspensor; secondary suspensors also occur in *Chamaecyparis* and *Cupressus*, but they are absent from *Thuja*. Cleavage polyembryony occurs in all the genera that have been studied, with the exception of *Thuja*. Embryogenesis in *Juniperus* presents some peculiar features: in a twelve-celled proembryo all the eight cells below the open 'u' tier elongate to give what is sometimes called a 'suspensor unit'. The lowermost cells then become lobed. Lobed and unlobed ends then intertwine to form a compact knot that bores its way into the gametophyte.[49] The tubes that have penetrated furthest then cut off terminal cells which give rise to multicellular embryos.

The basic chromosome number is remarkably uniform throughout the family ($n = 11$ in all the genera that have been examined).

Podocarpaceae

The Podocarpaceae, comprising seven genera and 150 species, is the most important family of conifers in the southern hemisphere. Although there are some representatives north of the equator, these are assumed to have arrived relatively recently, the family having originated in the south. The family is of great interest, not only because of its present geographical distribution pattern, but also because it shows a remarkably wide range of morphological characters. Thus, in some species the ovules are borne in cones, while in others the female cone is so reduced that it forms a swollen fleshy receptacle with a single terminal ovule. The male cones of a few species form part of a compound structure and are, therefore, at a very low level of evolution, lower even than that of the Lebachiaceae, for their closest parallel is to be found among the Cordaitales. Others have the simple male cone that is characteristic of all other conifers. Likewise, within this family there is a wide range of pollination mechanisms and there is considerable variation in the details of male gametophyte formation and of embryogenesis. Even the basic chromosome number ranges from 9 to 19, all intermediates except 14 and 16 being represented.

The genus *Podocarpus* is by far the largest in the family, with 110 species, and is represented in the western hemisphere (where it extends from Patagonia to the West Indies) and in the eastern (where it extends from New Zealand to Japan, with westward extensions into Burma). *Dacrydium* (22 spp.) and *Phyllocladus* (7 spp.) also have representatives north of the equator. The former occurs in New Zealand, Tasmania, Malaysia, Burma and south China, and is also found in Patagonia and Chile, while the latter is found chiefly in New Zealand, although represented also in Tasmania, New Guinea, North Borneo and the Philippines. *Saxegothaea* is a monotypic genus restricted to Chile; *Microcachrys* is also monotypic and is restricted to Tasmania, while *Microstrobos* (= *Pherosphaera*), with two species, occurs in eastern Australia, as well as in Tasmania. *Acmopyle* (3 spp.) is found in New Caledonia and the Fijian Islands.

Most of the Podocarpaceae bear their leaves and other parts in a spiral, but in *Microcachrys* they are decussately arranged. The ovule may be erect or (more usually) reflexed and, except in *Phyllocladus* and *Microstrobos*, the ovuliferous scale is more or less folded round the single ovule to form an extra envelope, called the 'epimatium'. This is either partially or completely fused to the integument and in some species is also fused to the bract-scale. The male cones bear numerous microsporophylls, each with two pollen-sacs, and the pollen-grains have either two or three air-bladders, although in some genera they are almost lacking. *Phyllocladus* (Fig. 32L) is a remarkable genus, not only because its photosynthetic organs are cladodes instead of leaves, but also because its ovules are partially enveloped in an 'aril' (Fig. 32N), a structure not found in any other conifers.

Podocarpus is not only the largest genus in the Podocarpaceae, it is also the largest genus of all present-day conifers. While most species are tall trees, some of which reach a height of 60 m, many are shrubs, and *P. ustus*, confined to New Caledonia, is remarkable in being the only conifer that is suspected of being parasitic. The leaves of this species are variously described as 'reddish bronze' or 'purple' and it is believed to be a partial parasite on the roots of *Dacrydium taxoides*.[80] In some species the leaves are up to 30 cm long and 5 cm wide, while in others they are minute and scale-like. Mostly, they are spirally arranged, but in a few species they are opposite or sub-opposite in two close ranks.

Starting in 1948, Buchholtz and Gray,[46] and subsequently Gray alone,[80] carried out a detailed revision of the taxonomy of the genus *Podocarpus*. Their conclusions indicate that the genus should be subdivided into eight sections, of which the most primitive is the *Stachycarpus* section, so called because its ovules are borne in cones. Figs. 32A and B illustrate the cone of *P. spicatus* at two stages of its growth and Fig. 32C shows how, in this species, the epimatium is fused throughout to the integument, whereas in *P. ferrugineus* (Fig. 32G) it is free, except at the chalazal end of the ovule. A male shoot of *P. spicatus* is illustrated in Fig. 32D, while Figs. 32E and F show how each male cone is borne in the axil of a bract, thus forming part of a compound structure. Very few species, however, show this primitive arrangement and even within the *Stachycarpus* section most species bear solitary male cones.

Podocarpus totara (Fig. 32H), from the section *Eupodocarpus*, shows considerable reduction in the female cone, for there are only two ovules in each, as in *P. macrophylla* too

(Fig. 32 I). Sometimes there is only one ovule in each cone, while in the *Dacrycarpus* section there is even greater reduction of the cone. Thus, Fig. 32J illustrates the way in which the ovules are borne singly on fleshy receptacles, and Fig. 32K shows something of the internal organisation of the ovule. In this species the bract-scale is completely fused to the epimatium. The lower parts of the cone swell up at maturity into the bright red succulent receptacle, which resembles a small plum. It is edible and is eaten by the Maoris, as well as by birds.

Some species of *Dacrydium* are tall trees up to 30 m high, while others are only shrubs. The leaves of most species are scale-like, but those of *D. taxoides* are always of the juvenile linear type. *D. Franklinii* is like *Podocarpus spicatus* in having spike-like female cones. Other species, however, have much reduced cones, there being just a single terminal ovule, partially enclosed by a cup-shaped structure which is sometimes incorrectly called an aril, but which in fact is formed from sterile cone-scales. There is a similar structure which partially envelops the seed of *Phyllocladus* (Fig. 32N), but in this genus there is a true aril as well. This is a secondary structure, without any vascular supply, which grows up round the integument, but which remains free from it (Figs. 32M and N).

The range of pollination mechanisms within the Podocarpaceae is as wide as that found in the Pinaceae. At one extreme, *Podocarpus* has a mechanism similar to that of *Pinus*, the pollen-grains having two air-bladders which cause them to float up towards the nucellus in a copious fluid exudate. At the other extreme, *Saxegothaea* is comparable with the most advanced type of *Tsuga*, where there is no exudate, the pollen-grains are without air-bladders and alight anywhere on the cone-scale. The nucellus projects as a marked bulge beyond the micropyle (Fig. 32 O); the pollen-tube grows into it and there remains dormant for some nine months. The pollen of *Phyllocladus* has only rudimentary air-bladders and the micropyle, at the time of pollination,

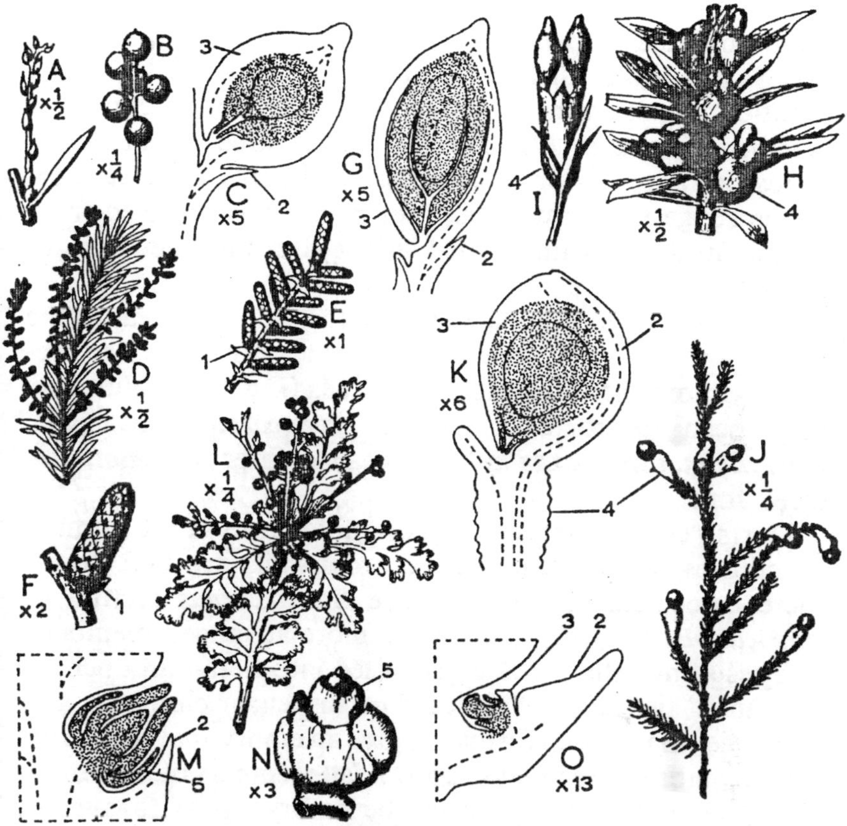

FIG. 32

Podocarpaceae

Podocarpus spicatus: A, young female cone; B, mature female cone; C, l.s. ovule; D, shoot with male cones; E, male reproductive branch, showing its compound nature, with cones each in the axil of a bract; F, single male cone and subtending bract. *P. ferrugineus:* G, l.s. ovule. *P. totara:* H, female shoot with fleshy receptacle, and paired seeds. *P. macrophylla:* I, young female cone with two ovules. *P. dacrydioides*: J, female shoot with fleshy receptacles and single seeds; K, l.s. young cone with single ovule. *Phyllocladus:* L, female shoot of *P. trichomanoides*; M, l.s. ovule and part of female cone of *P. glaucus*; N, ovule of *P. alpinus*, showing aril. *Saxegothaea:* O, l.s. cone-scale of *S. conspicua.* (1, bract; 2, bract-scale; 3, epimatium = ovuliferous scale; 4, receptacle = cone-axis; 5, aril)

(A, B, J, L, N, after Polle and Adams; C, G, K, M, Sinnott; D, F, H, Hooker; E, Wilde; I, Eicher; O, Hirmer)

points obliquely upwards. However, apart from this genus and *Saxegothaea*, all Podocarpaceae are believed to have a flotation mechanism. The fact that the pollen-grains of *Microcachrys* and *Microstrobos* have three air-bladders, instead of two, is worthy of note, for it lends some support to the suggestion that both types might have evolved from an ancestral type with an air-bladder all round the grain, as in the Cordaitales.

There are strong similarities with the Araucariaceae in the development of the male gametophyte, for there are frequently more than two prothallial cells (Fig. 30A). Numbers have been recorded as follows: *Podocarpus*, one–eight; *Dacrydium*, three–six; *Phyllocladus*, one–three; *Saxegothaea*, three–four.

During embryogenesis there are differences in the number of mitoses which occur before cross-walls are laid down, there being five in the primitive *Stachycarpus* section of *Podocarpus*, but only four in the more advanced members. The resulting cells become organised into tiers—an open 'u' tier, an 's' tier and an 'e' tier. The suspensor elongates and the 'e' tier may divide further to give an embryonal suspensor and an embryo. As far as is known, all members of the family are alike in that the cells of the embryo pass through a binucleate phase before they revert to a uninucleate condition, through what is called an embryo-tetrad stage, where the embryo consists of tetrads of cells. In the more primitive members these tetrads remain together and cleavage does not take place. In the more advanced members, however, cleavage polyembryony occurs—a fact which has an important bearing on the assessment of the evolutionary status of polyembryony in the conifers as a whole.

As already mentioned, there is a wide range of basic chromosome numbers within the family, but there is also a wide range even within some genera. Thus, the haploid numbers in *Dacrydium* are 9, 10, 11, 12 and 15, while in *Podocarpus* they are 10, 11, 12, 13, 17, 18 and 19. It is perhaps significant that the two highest numbers occur in the primitive *Stachycarpus* section of *Podocarpus*. In *Phyllocladus* the number is 9, and in *Saxegothaea* it is 12.

Cephalotaxaceae

This family is represented today by the one genus *Cephalotaxus* (6 spp.) restricted to eastern Asia, but in past ages there were representatives in Europe and western U.S.A. also. Modern species of *Cephalotaxus* are shrubs or small trees up to 15 m high, with opposite or whorled branches, bearing spirally arranged linear leaves up to 7 cm long. At one time the genus was thought to be related to *Torreya*, a member of the Taxales, and it is true that the secondary wood does show a close resemblance to that of the Taxales, for there are abundant tertiary spiral thickenings on the walls of the tracheids. However, its ovules are borne in cones instead of terminally.

All six species are dioecious. The female cones have a few opposite and decussate bracts, each subtending a pair of erect ovules on an extremely abbreviated secondary fertile short shoot (Figs. 31Q and R). Normally only one ovule develops into a large olive-like seed, up to 3 cm long (Fig. 31P), with an outer fleshy layer surrounding a stony layer. The integument is fused with the nucellus, except at the apex, and there are two vascular bundles in the outer fleshy layer, corresponding with the position of the two opposite primordia that give rise to the integument. The male cone is made up of spirally arranged microsporophylls, each with two or three pollen-sacs. There are no air-bladders on the pollen-grains and pollination involves a simple drop-mechanism. Male prothallial cells are completely lacking (Fig. 30B), and at pollination the pollen-grain contains merely a tube-nucleus and a generative nucleus. It remains in this condition throughout the winter, for fertilisation does not occur until the season following pollination. There seems to be some disagreement as to whether the gametes are of unequal size, but the archegonia are widely separated and each pollen-tube penetrates only one.

During embryogenesis, wall-formation occurs at the sixteen-cell stage. There is an open 'u' tier of three–five cells, a suspensor of three–five cells and eight–ten 'e' cells, of which the lowermost form a cap which soon disintegrates. As the suspensor elongates, some of its cells give rise to 'suspensor embryos', while the uppermost 'e' cells give rise to a secondary suspensor. Cleavage rarely occurs.

The haploid chromosome number is 12.

Araucariaceae

This family is an extremely old one, for its fossil record extends back to the Triassic, when it was represented in both the northern and the southern hemispheres. Today, however, there are just two genera, which are restricted to the southern hemisphere: *Agathis* (Kauri Pines—20 spp.) and *Araucaria* (16 spp., of which *A. araucana* is the familiar Monkey Puzzle and *A. heterophylla* is the Norfolk Island Pine). *Agathis* is exclusively eastern, extending from the Philippines to New Zealand and from Malaya to Fiji, whereas *Araucaria* occurs in South America, as well as in Australia, New Guinea and New Caledonia.

Mention has already been made of the leaves of the Araucariaceae, which may be relatively large and may be retained for many years, the leaf-bases expanding as the branch bearing them enlarges. Not only are the leaves spirally arranged, but so also are the other appendages, although the main branches are usually whorled, as in *Lebachia*. *Agathis* is usually monoecious and *Araucaria* dioecious. The female cones are almost spherical, and those of *Araucaria Bidwillii* may be as much as 30 cm in diameter, but those of *Agathis australis* (Fig. 33A) seldom exceed 6 cm. The ovuliferous scale is intimately fused to the bract-scale, except in *Araucaria*, where the tip of the ovuliferous scale is free and constitutes the so-called 'ligule' (Figs. 33D–F). There is normally only one reflexed ovule on each cone-scale.

The male cones of *Araucaria Rulei* may exceed 20 cm in length and each microsporophyll bears many pollen-sacs. There are five–fifteen in *Agathis* and eight–fifteen in *Araucaria* (Figs. 33H and I). From such a cone the output of pollen-grains might well be as high as 10,000,000, according to Chamberlain.[5] The pollen-grains are without air-bladders, and are interesting for several reasons. Large numbers of extra prothallial cells are produced (Figs. 30A and 33G); thirteen–forty in *Araucaria* and six–ten in *Agathis*. By contrast, the pollination-mechanism must be regarded as relatively advanced. There is no drop-mechanism at all; instead, the pollen lodges on the cone-scale (or, in *Araucaria*, on the ligule) and the pollen-tube grows towards the ovule, whose nucellus has a beak projecting through the micropyle. There is some controversy as to whether the male gametes are unequal in size, but, in view of the fact that there is only one archegonium in each ovule, only one male gamete can function.

In their embryogenesis the Araucariaceae differ markedly from all other conifers. Free-nuclear divisions of the zygote take place until there are about thirty-two nuclei in *Araucaria*, or up to sixty-four in *Agathis*, and they are peculiar in being grouped together in the centre of the egg-cell. Cell-walls are then laid down and the cells are arranged in such a way that there is a central group with two tiers surrounded by an outer jacket (Fig. 33B). The upper cells of the jacket elongate and become the suspensor, while the lower jacket-cells form a cap of unknown function (Fig. 33C). The suspensor cells divide once and, as they elongate, they may even project through the apex of the archegonium in *Agathis*. The upper of the two central tiers of cells gives rise to the secondary suspensor, and the lower the embryo, there being no cleavage polyembryony at all. Neither, of course, does simple polyembryony occur, for there is only a single archegonium in each ovule.

The basic haploid chromosome number in the Araucariaceae is 13.

Apart from the brief statement, near the beginning of this chapter, that each family of conifers shows some advanced characters and some primitive ones, no attempt has been made, so far, to assess their relative advancement. It is not easy, however, to be more precise, for, unfortunately, there is an arbitrary element in the decision as to which characters should be taken as the basis for any more accurate assessment. Those about which there is the least dispute, as indicators of primitiveness, are (1) a well-developed female cone, (2) bract-scale free from the ovuliferous scale, (3) winged pollen, (4) male prothallial cells numerous, (5) shoot-apex with mother-cells, but (6) without separation into

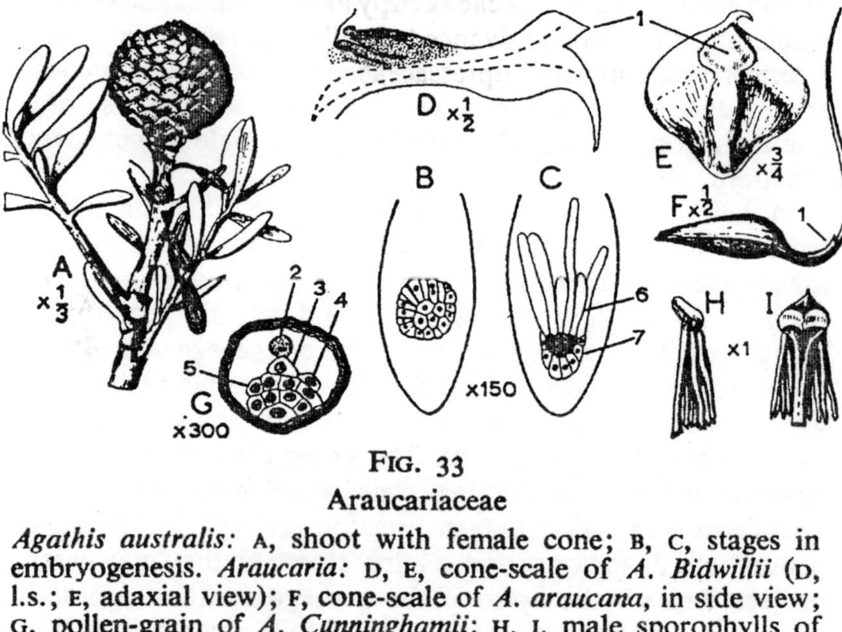

FIG. 33

Araucariaceae

Agathis australis: A, shoot with female cone; B, C, stages in embryogenesis. *Araucaria:* D, E, cone-scale of *A. Bidwillii* (D, l.s.; E, adaxial view); F, cone-scale of *A. araucana*, in side view; G, pollen-grain of *A. Cunninghamii*; H, I, male sporophylls of *A. angustifolia* (H, side view; I, adaxial view)

(1, ligule = ovuliferous scale; 2, tube-nucleus; 3, spermatogenous cell; 4, sterile cell; 5, prothallial cells; 6, suspensor cells; 7, cap-cells)

(A, after Poole and Adams; B, C, Eames, adapted by Chowdhuri; D, Hirmer; E, F, Pilger; G, Chamberlain; H, I, Eichler) tunica and corpus, (7) prolonged free-nuclear phase in early embryogenesis, (8) absence of cleavage during later embryogenesis (despite the contrary belief of Buchholtz). A high chromosome number is held by some to be correlated with general primitiveness and homogeneous wood-rays are regarded by Greguss as being more primitive than heterogeneous.

On the basis of these characters it is possible to make a rough assessment of relative primitiveness, and the most primitive family appears to be the Podocarpaceae. However, there is a wide range within this family, the *Stachycarpus* section being extremely primitive, while the *Dacrycarpus* section is clearly more advanced. The Pinaceae, Cephalotaxaceae and Araucariaceae occupy an intermediate position while the two most advanced families appear to be the Taxodiaceae and the Cupressaceae.

9

Taxales

Profusely branching, evergreen shrubs or small trees, with spirally arranged small linear leaves. Wood pycnoxylic; tracheids with abundant tertiary spirals. No resin canals in wood or leaves. Ovules solitary, arillate, terminating a dwarf shoot, with decussate bracts. Microsporangiophores in small cones (in *Austrotaxus*, each sporangiophore subtended by a bract), scale-like or peltate, with two–eight pollen-sacs. Embryo with two cotyledons.

Taxaceae

*Palaeotaxus**, *Taxus*, *Austrotaxus*, *Pseudotaxus* (= *Nothotaxus*), *Torreya*, *Amentotaxus*

Until early in the present century this small group. comprising only five living genera and about twenty species, was usually classed among the conifers, along with *Cephalotaxus* and the podocarps. Then, in 1920, Sahni[148b] suggested that *Taxus*, *Torreya* and *Cephalotaxus* should be placed in a phylum of their own—Taxales. Florin[65, 67] agrees with this, except for *Cephalotaxus* which he regards as a true conifer. The basis for this separation of the Taxales from the Coniferales is the discovery of fertile specimens of *Palaeotaxus* in the Triassic, and of *Taxus jurassica* in Jurassic rocks of Yorkshire, both of which bore solitary terminal ovules. From this it is concluded that, whereas the solitary ovules of some podocarps represent the ultimate stage in reduction of a cone, those of taxads are fundamentally solitary and terminal. Some taxonomists make an even greater separation into Taxopsida. This, however, seems to be going too far, for it overlooks the many features which the Taxales share with the Coniferales.

The seed of *Palaeotaxus* is illustrated in Fig. 34A, which shows how it was borne at the tip of a short shoot and was partially enclosed by spirally arranged bracts. It also shows how the aril (stippled) enclosed the ovule, except right at the micropylar end.

Taxus, represented by nine species, occurs in North America, Europe and Asia, and even extends into Malaysia. The fact that the seeds are distributed by birds is held to explain, in part, this wide geographical range. *Taxus baccata*, the yew, is capable of growing to a height of nearly 20 m, with a massive trunk 7 m or more in girth. Its leaves are linear, 2–3 cm long, and are spirally arranged, except for the scale-like leaves on the fertile short shoots, which are opposite and decussate.

The zonation of the apical meristem is generally like that of the Taxodiaceae and the Pinaceae, except that the frequency of periclinal divisions in all regions varies considerably as the growing season advances. At no time is there a distinct tunica. The secondary wood, formed from a single persistent cambium, is dense, being completely without wood-parenchyma. It is made up of tracheids with uniseriate pits and abundant tertiary spirals. The latter are most noticeable and may, perhaps, explain the extraordinary elasticity of yew-wood and its use in the making of bows for archery. The wood-rays are homogeneous and are composed of heavily thickened cells.

The majority of yew-trees are unisexual, but occasional monoecious specimens are known to occur. Fig. 34G illustrates the way in which the ovuliferous short shoot stands in relation to the apex of the main shoot, which later overtops the short shoot. At this stage the aril is still in a primordial state, and it remains so until after fertilisation has taken place and an embryo has begun to develop (Fig. 34D). Ultimately, however, it extends beyond the seed, as a red juicy cup, attractive to birds (Figs. 34B and C). The integument is fused to the nucellus, except at the apex, and becomes stony. There are usually two faint vascular bundles, represented by strands of phloem only, running up inside this stony layer. The xylem of these strands extends no further than the chalaza.[57] Their arrangement, alternating with the uppermost pair of scale-leaves, corresponds with that of the two primordia from which the integument arises. The aril, likewise, receives two vascular bundles, but they are minute and rudimentary.

At meiosis, a single row of megaspores is produced, of which several may develop into female prothalli. Free-nuclear divisions occur until about 256 nuclei have been formed, and then cell-walls are laid down. As many as twenty-five archegonia may appear, though the normal number is about ten. Each consists of two to four neck-cells and an egg-nucleus, there being no ventral-canal-nucleus.

One of the most remarkable features of *Taxus* is its peltate microsporangiophores (Fig. 34E) which are strongly reminiscent of *Equisetum* in having six–eight reflexed sporangia. They are arranged in small cones (Fig. 34F), terminating short shoots, and yield vast clouds of pollen-grains in early spring. There are no male prothallial cells, and the pollen-grains contain just a single nucleus when they are shed. Pollination involves a drop-mechanism and occurs at about the time that meiosis is taking place in the ovule. Subsequently, a tube-nucleus, a sterile cell and a spermatogenous cell are formed, and eventually two male cells, one of which is much larger than the other. The pollen-tubes spread out in a characteristic way, to form a sac-like expansion, which may cover the entire apex of the female prothallus.

After fertilisation up to thirty-two nuclei are formed from the zygote, before wall-formation occurs to give nine to thirteen open 'u' cells, nine to thirteen suspensor 's' cells and six to fourteen embryo 'e' cells. Cleavage polyembryony rarely occurs.

Pseudotaxus (= *Nothotaxus* 1 sp.) occurs in a small region of east China, while *Austrotaxus* (1 sp.) is restricted to New Caledonia. The latter is remarkable for its male cones, on

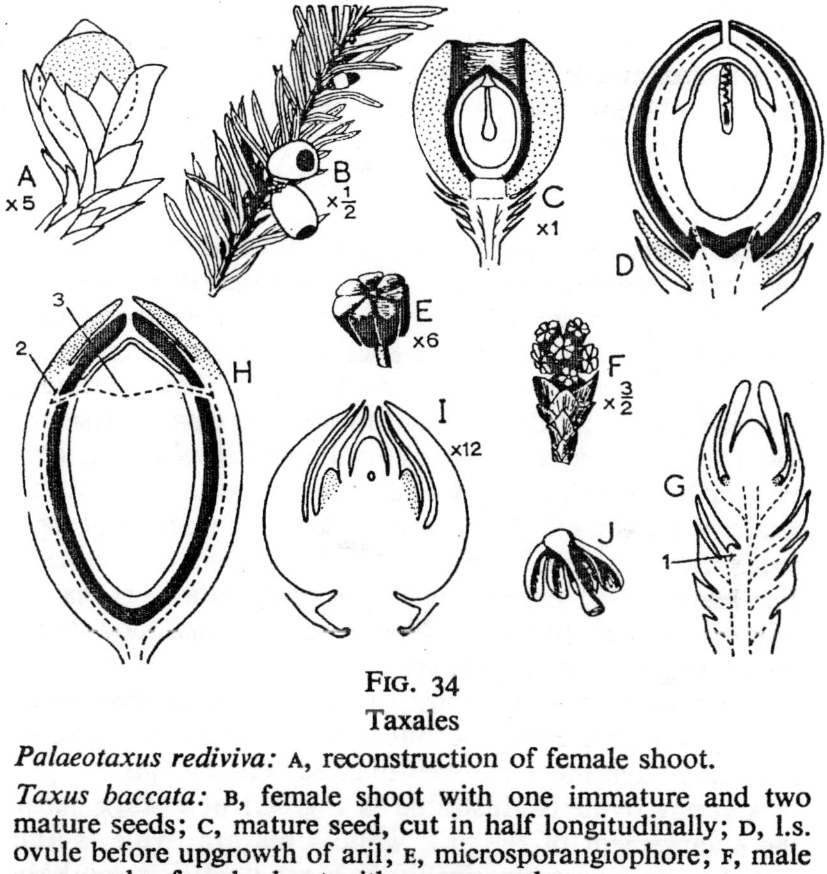

FIG. 34
Taxales

Palaeotaxus rediviva: A, reconstruction of female shoot.

Taxus baccata: B, female shoot with one immature and two mature seeds; C, mature seed, cut in half longitudinally; D, l.s. ovule before upgrowth of aril; E, microsporangiophore; F, male cone; G, l.s. female shoot with young ovule.

Torreya taxifolia: H, l.s. seed (diagrammatic); I, l.s. young ovule, before intercalary growth; J, microsporangiophore.

(1, apex of main axis; 2, foramen; 3, vascular loop; aril-tissue stippled)

(A, after Florin, B, C, E, F, Beissner; D, Sahni; G, Hirmer; H, Oliver; I, Coulter and Land; J, Hooker)

the basis of which Nakai[133] suggested the creation of a separate family—Austrotaxaceae. These are made up of sporangiophores, each subtended by a bract.[151] The sporangiophores terminate in a synangium of three or four fused sporangia, and look remarkably like the sporangial apparatus of the pteridophyte *Psilotum nudum*. *Pseudotaxus* also has bracts in the male cone, but they are irregularly distributed. It is difficult to reconcile this arrangement with that in other taxads or, indeed, with that in any of the conifers. Wilde[199] attempted to do so by supposing that each so-called 'sporangiophore' is in effect a greatly reduced male cone, the whole structure being regarded as a compound structure, but it must be admitted that there is very little real evidence for this ingenious interpretation. Florin[65] draws a parallel with *Cordaitanthus Penjonii*, where the sterile and fertile appendages are not clearly segregated.

Torreya (5 spp.) has a very disjunct distribution pattern, occurring only in California, Florida and eastern Asia. Its fossil remains, extending back to the middle Jurassic, however, occurred in Yorkshire and Europe as well. Some species attain a height of 30 m, with a girth of 3 m, but most specimens are usually much smaller—often only shrubs. The leaves of *T. californica* (Californian Nutmeg) may be as much as 8 cm long, but, like those of *Taxus*, they receive only a single vascular bundle.

The secondary wood differs from that of *Taxus* in containing wood-parenchyma, but is otherwise similar, and the tracheids have the same well-marked tertiary spirals.

A young ovule is illustrated in Fig. 34I, at a stage before the enlargement of the aril, while a mature ovule is illustrated (very diagrammatically) in Fig. 34H. It is remarkable in having the aril fused to the integument over most of its length. Its vascular supply is even more remarkable, for, although the integument lacks vascular bundles of its own, there are two which run up inside the aril nearly to the apex of the seed, and then each sends a branch through a foramen in the stony layer; each branch then forks, to give a loop which encircles the seed, just beneath its apex. This extraordinary arrangement is explained by Oliver,[135] by supposing that the two foramina are morphologically at the chalazal end of the seed, but that intercalary growth has resulted in their displacement upwards. This would also explain the 'fusion' of the aril to the integument. In the mature seed the female prothallus grows vigorously, but irregularly, into the surrounding tissues, with the result that its outline becomes very complicated and 'ruminate' (not shown in Fig. 34H).

The male organs, instead of being peltate, as in *Taxus*, are dorsiventral scale-like structures, with four abaxial pendulous pollen-sacs (Fig. 34J). As in *Taxus*, no male prothallial cells are formed, and the male gamete cells are very unequal in size.

Only a single archegonium is formed in each female prothallus, and there is no ventral-canal-nucleus (as in *Taxus*).

Amentotaxus (4 species) is confined to east Asia at the present day, but fossil remains have been described from Europe and western America, too. It is remarkable for the way in which the seed, with its aril and enclosing bracts, is borne at the tip of a peduncle up to 1·5 cm long. Yamamoto[200b] considers that it should be placed in a separate family—Amentotaxaceae—but it shows sufficient similarities with *Torreya* to persuade most morphologists otherwise. Thus, its aril is fused to the integument and, whereas in *Taxus* the vascular supply to the integument is much reduced, in *Torreya* and *Amentotaxus* it is completely lacking—a condition which Florin associates with the marked intercalary growth of the 'hyposperm' in these two genera.

10

Ginkgoales

Branching trees with long and short shoots (except in the earliest fossil members). Wood pycnoxylic. Leaves leathery, strap-shaped or fan-shaped, often deeply divided, with dichotomous venation. Ovules two to ten, terminal on axillary, branching or almost unbranched, axes. Seed large, with fleshy outer layer and stony middle layer. Male organs axillary, unbranched, catkin-like, bearing microsporangiophores each with two to twelve pendulous microsporangia. Sperm with spiral band of flagella.

Trichopityaceae*
Trichopitys

Ginkgoaceae
*Sphenobaiera**, *Ginkgoites**, *Baiera**, *Arctobaiera**, *Windwardia**, *Eretmophyllum**, *Ginkgo*

This group of gymnosperms, comprising many species and at least sixteen genera, was once almost world-wide in its distribution, but is now represented by the single species *Ginkgo biloba*, which, if it occurs naturally anywhere, is restricted to a small and relatively inaccessible region in south China. These facts alone would suffice to explain the interest of the single surviving species, but there is the additional fact that leaves identical with those of *Ginkgo biloba* occurred as far back as the Triassic some 200,000,000 years ago. Such remarkable antiquity has captured the imagination of botanists and laymen alike, for, as Seward[159] remarked, we see *Ginkgo biloba* 'as an emblem of changelessness, a heritage from worlds of an age too remote for our human intelligence to grasp, a tree which has in its keeping the secrets of the immeasurable past'.

Trichopitys heteromorpha, from the Lower Permian of southern France, has been known since 1875, and, from the first, was suspected of having ginkgoalean affinities. Recent investigations have confirmed this, and Florin[66] concludes that it was the earliest member of the group. However, it differed in several important respects from other members and is, therefore, regarded here as properly belonging to a separate family—Trichopityaceae. A portion of the fertile shoot is illustrated in Fig. 35A, which shows how the ovules were borne on branching structures subtended by dichotomous leaves. There are indications that the leaves, instead of being flattened, were circular in cross-section, and that they branched in several planes. The ovules (Fig. 35B), unlike those of *Ginkgo*, were reflexed and showed a striking similarity to those of *Cordaitanthus pseudofluitans* (Fig. 25D).

It is not known what the male organs of *Trichopitys* were like, but *Sphenobaiera furcata*, of Triassic age, bore clusters of microsporangia at the branch-tips of a bifurcating axis. These, in turn, were borne on short shoots, along with leaves that were very like those of *Trichopitys*.[112] Most of the remaining fossil genera are known only from their leaves, but associated with those of the Jurassic *Baiera muensteriana* were some catkin-like structures whose short lateral branches terminated in ten to twelve reflexed microsporangia (Figs. 35D and E). Fig. 35C shows how deeply divided were the leaves of *Baiera*, all the segments being in one plane. So variable were the leaves of most species, however, that it is difficult to distinguish them from each other or from those of *Ginkgoites* (Fig. 35I), many species of which are known, some deeply lobed, others scarcely so.

The leaves of *Baiera*, *Ginkgoidium* and *Ginkgoites* were like those of modern *Gingko* in having a distinct petiole.

The other fossil genera, which lacked this feature, are separated from each other by such characters as overall shape, depth of lobing, number of veins in the lamina, distribution of stomata (i.e. whether amphistomatic or hypostomatic) and the structure of the stomatal apparatus itself. Thus, some genera had strap-shaped leaves with few veins, e.g. *Arctobaiera* (Fig. 35F) and *Windwardia* (Fig. 35G), while others had fan-shaped leaves with many veins, e.g. *Eretmophyllum*. Whatever the overall shape, however, all had strictly dichotomous open venation, as do the leaves of *Ginkgo biloba* today. In growth habit, too, the fossil Ginkgoales (with the exception of *Trichopitys*) were like *Ginkgo biloba* in having long and short shoots.

Until 1951 the Jurassic genus *Czekanowskia* was widely believed to belong to the Ginkgoales. Its leaves were borne on short shoots and dichotomised three or four times into linear segments. Harris,[97] however, has shown that they were associated with some remarkable reproductive structures, known as *Leptostrobus*, which were certainly quite unlike those of Ginkgoales and, indeed, were unique in the plant kingdom. They consisted of a loose cone with bivalved capsules, like the two halves of a cockle-shell, each half bearing three to five seeds. Their association together in the rocks of Greenland, Yorkshire and western Siberia makes it almost certain that these leaves and reproductive organs belonged to the same plant. Several palaeobotanists have recommended that a new group should be created—Czekanowskiales—to accommodate this extraordinary seed-plant. So far as the fossil Ginkgoales are concerned, however, this should serve as a warning against putting too much reliance on statements about genera that are based solely on leaves.

Ginkgo biloba, sometimes called the Maidenhair Tree, is widely cultivated in China and Japan for its edible seeds. When young, its growth habit is strangely spindly, but older specimens acquire a broad spreading crown and may attain a height of 30 m. The leaves (Fig. 35H) are fan-shaped, with beautifully regular dichotomous venation, and have a petiole which receives two endarch vascular strands. Those on the long shoots are mostly bilobed, but those on the short are entire (Figs. 35J and O). In autumn they change colour to a golden yellow before being shed. Until recently it has always been stated that the leaves are hypostomatic. However, Kanis and Karstens[105] have recently shown that this is not entirely true, since the leaves from long shoots (of male trees in particular) tend to have a few stomata in the upper epidermis. The stomata in the lower epidermis occur in broad bands between the veins and are surrounded by four, five or six accessory cells with finger-like processes projecting over the guard-cells.

Short shoots may frequently become more active and develop into long shoots or, more rarely, long shoots may terminate in short ones. This suggests that there is no very fundamental difference between them, and the fact that the apical meristems of the two kinds of shoot are essentially the same is in keeping with this. Because of occasional periclinal cell-divisions in the outer layer, there is no distinct tunica. From a region of relatively inactive large central mother-cells, the more active cells of the peripheral region and the rib-meristem are produced, and it is in the relative rates of cell-division in the latter region that the difference between long and short shoots is said to lie. These different rates are in turn said to be influenced by auxin concentrations.[82]

The primary xylem of the stem consists of a number of separate strands which branch sympodially when the leaf-traces are given off, and it is an interesting fact that the two traces to any one leaf arise quite independently from two different primary strands.[81] Since there are no anastomoses between the veins in the lamina, the two halves of the leaf are, therefore, completely isolated from each other. The secondary wood is formed from a single cambium and is pycnoxylic, being made up of tracheids, and wood-rays that are seldom more than one cell wide. The tracheids have one or two rows of bordered pits on the radial walls. There are abundant mucilage canals throughout the plant—in the pith and the cortex of the stem, and even between the veins of the leaves.

Ginkgo biloba is dioecious, and the sex is determined by

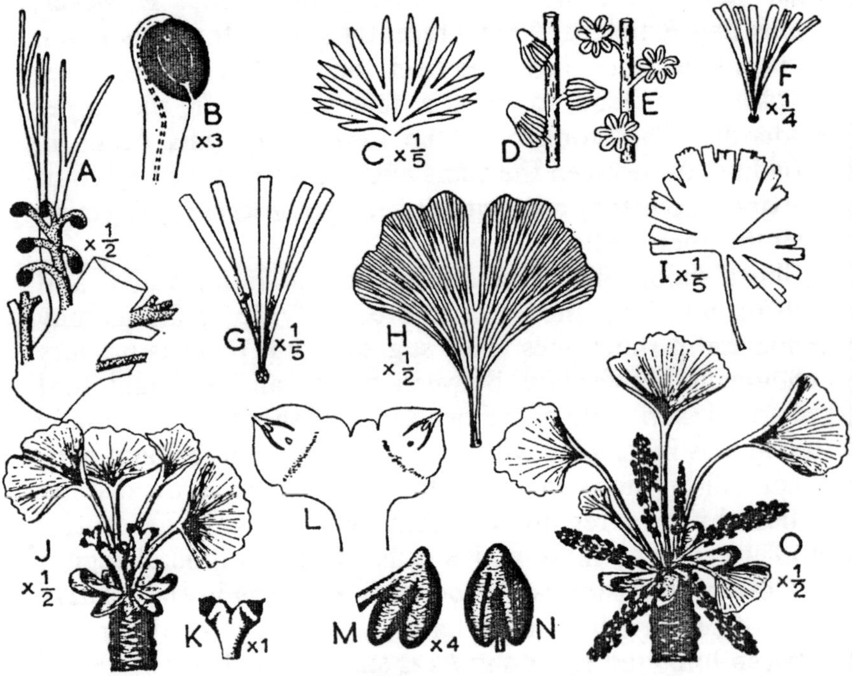

FIG. 35

Ginkgoales

Trichopitys heteromorpha: A, ovuliferous short shoot in the axil of a dichotomous leaf; B, reflexed ovule.

Various fossil leaves: C, *Baiera* sp.; F, *Arctobaiera Flettii*; G, *Windwardia Crookallii*; I, *Ginkgoites acosmia*.

Microsporangia of *Baiera muensteriana*: D, immature; E, at pollination. *Ginkgo biloba*: H, leaf from a long shoot; J, short shoot with young ovules; K, pair of young ovules; M, N, microsporangia (N, dehiscing); O, short shoot with microsporangia.

(A, B, F, G, after Florin; C, Knowlton; D, E, Schimper-Schenk; H, Chamberlain; I, Harris; J, K, O, Wettstein; L, Sprecher)

sex chromosomes, the female plant being of the XX type and the male XY.[119]

The ovules are borne on peduncles that arise in the axils of leaves or scale-leaves on the short shoots (Fig. 35J). The tip of the peduncle bifurcates, usually once only, and each branch bears a single sessile ovule with a fleshy collar round its base (Figs. 35J–L). Usually there are just two ovules on each peduncle, but, occasionally, there may be three, four or more; and however many there are, there are twice as many vascular bundles in the peduncle. This fact has often been mentioned in support of the idea that each ovule is borne on a sporophyll, and the collar has been regarded, by some, as a modified lamina. In further support, teratological freaks have been described, in which the ovule is associated with a leaf-like lamina. However, now that the classical theory of plant organisation has largely been abandoned, it is no longer necessary to force a sporophyll interpretation on to all reproductive organs, and it is possible to accept the peduncle of *Ginkgo* for what it is —a stem structure with terminal ovules.

In the young ovule the integument appears to be free from the nucellus (Fig. 35L), but subsequent intercalary growth results in enlargement of the chalazal end, so that in the mature ovule the integument is 'fused' to the nucellus, except at the apex (Fig. 36G). Meiosis gives a linear tetrad of spores, the lowermost of which gives rise to the female prothallus. At about the time of pollination, in spring, free-nuclear divisions occur (Fig. 36B) until about 256 free nuclei have been produced, and then cross-walls are laid down. Two or three archegonia appear, each consisting of a spherical egg-cell, a ventral-canal-cell and a neck of four cells (Fig. 36C). The gametophyte is remarkable in containing abundant chlorophyll.

The catkin-like male organs are borne in the axils of leaves or scale-leaves, on the short shoots of male trees (Fig. 35O), and consist of a main axis with lateral microsporangiophores. These have a terminal knob, containing a mucilage sac, and they bear two pendulous microsporangia which dehisce by means of a longitudinal slit. Some morphologists have suggested that the terminal knob represents an abortive sporangium, holding that the sporangiophore of *Ginkgo* has evolved by reduction from one with many sporangia, as in *Baiera* (Figs. 35D and E).

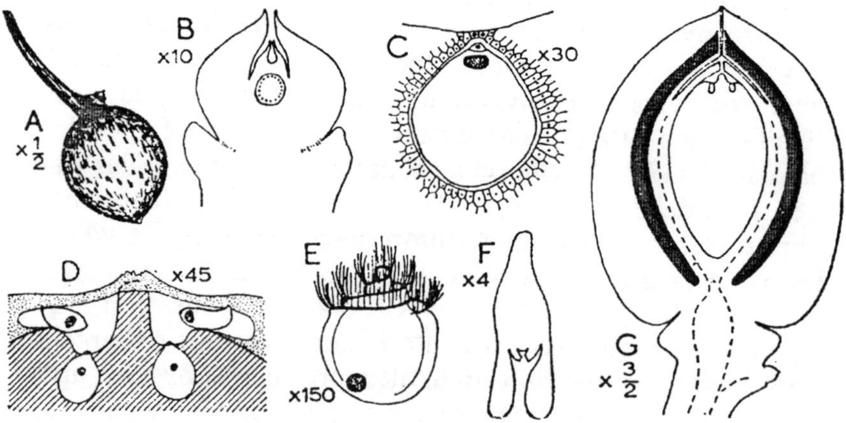

FIG. 36

Ginkgoales (continued)

Ginkgo biloba: A, ripe seed; B, l.s. ovule at the stage of free-nuclear division in the female prothallus; C, l.s. archegonium; D, l.s. archegonial chamber; E, spermatozoid; F, l.s. embryo; G, l.s. mature ovule (diagrammatic).

(A, after Wettstein; B, Coulter and Chamberlain; C, Strasburger; D, Hirasé; E, Shimamura; F, Sprecher)

Development of the male prothallus is essentially the same as in *Pinus*, with the formation of two prothallial cells, a tube-nucleus, a sterile cell and a spermatogenous cell.

Within the ovule there is a well-developed tent-pole which lifts the nucellar cap and, thereby, enlarges the pollen-chamber (Fig. 36D). The pollen-tube, although very limited in its growth compared with that in Cycads, is, nevertheless, haustorial and serves to anchor the pollen-grain to the side of the pollen-chamber, into which are released two large motile sperms, each with a spiral band of flagella (Fig.

36E).[161] By this time—autumn—the ovule has grown to full size (Figs. 36A and G) and may even have fallen to the ground before fertilisation has occurred. The outer layer of the seed is fleshy and gives off a nauseous odour of rancid butter. The middle layer is stony, while the inner layer is watery and contains a mantle of faintly lignified iso-diametric tracheids. The edible part is the female prothallus, which is eaten roasted.

As in most gymnosperms, there is a period of free-nuclear divisions after zygote-formation, but there is no well-defined suspensor. The embryo usually has two cotyledons (Fig. 36F) which, unlike the foliage leaves, have mesarch vascular traces.

The haploid number of chromosomes in *Ginkgo biloba* is 12.

There have long been arguments as to the true affinities of the Ginkgoales, some believing them to be related to the Cycads, with which they share the primitive feature—motile sperms. However, the work of Florin now leads to the conclusion that their affinities lie, rather, with the Coniferopsida, even though their relationship to other members of this group must be somewhat remote. Florin's conclusion is that the Ginkgoales, the Cordaitales, the Coniferales and the Taxales belong to the same natural group, the Coniferopsida, but that they constitute parallel evolutionary lines which were probably already separated in Upper Devonian or Carboniferous times.

11

Gnetales

Woody plants; trees, shrubs, lianes or stumpy turnip-like plants with stem partly below ground. Leaves opposite or whorled, simple, broadly elliptic or strap-shaped or reduced to minute scales. Secondary wood with vessels. 'Flowers' unisexual and normally dioecious (except in some monoecious species of *Gnetum*). Flowers organised into compound strobili or 'inflorescences'. Female flowers with a single erect ovule, the nucellus of which is surrounded by two to three envelopes, the micropyle projecting as a long tube. Male flowers with a 'perianth' and antherophores with one to eight synangia. Fertilisation by means of a pollen-tube with two male nuclei. Embryo with two cotyledons.

Gnetaceae

Gnetum

Welwitschiaceae

Welwitschia

Ephedraceae

Ephedra

Despite the fact that there are only three genera in the Gnetales, the volume of literature about this fascinating group is enormous, and even as early as 1912 studies of the group were described as 'immeasurable'. The main reason for this is that the three genera approach more nearly to the angiosperms than do any other gymnosperms. Indeed, many botanists in the past believed the Gnetales to be the ancestors of flowering plants, and even now there are still some who maintain that there is a close affinity between *Ephedra* and *Casuarina*.

The most noteworthy features that are common to the three genera are: (1) the presence of vessels in their secondary wood, (2) the arrangement of the 'flowers' in compound strobili, (3) the presence of a 'perianth' in the male flowers and of several envelopes (variously interpreted as perianth or extra integuments) round the ovules, (4) the great length of the micropyle, which projects as a long bristle-like tube. In other respects the three genera are so dissimilar that some taxonomists place them in three distinct orders.[58] Whether they should be so treated, or whether they should be placed in a single order, is, however, a matter entirely of personal preference, for they are completely isolated, with no fossil record at all, except for remains of *Ephedra* pollen from the Eocene, and some pollen rather like that of *Ephedra* and of *Welwitschia* from the Permian.[8]

The feature, above all others, that seems to have caught the imagination of the early morphologists is the presence of vessels (Figs. 37B–G). Those of *Welwitschia* (Fig. 37G) and the majority of those found in *Gnetum* (Fig. 37D) do, indeed, look very like an advanced type of angiosperm vessel, but it is now realised that this resemblance is no more than a remarkable example of convergent evolution. It is generally accepted that angiosperm vessels have evolved from tracheids with scalariform thickenings. Gnetalean vessels, by contrast, have evolved from pitted tracheids, and support for this view is provided by the occurrence of vessels that show intermediate stages between pits and perforations, or between many perforations and a single perforation. Thus, Fig. 37E shows an end-plate of *Ephedra major*, in which there are both bordered pits and perforations, while Fig. 37F illustrates a typical multi-perforate vessel in *E. californica*.

Figs. 37B–D are of different vessels from *Gnetum africanum*, showing varying degrees of fusion of adjacent perforations. It is interesting to reflect that it was the existence of vessels that first suggested a possible affinity between the Gnetales and the angiosperms, and that now it is a detailed knowledge of their vessels which suggests that no such affinity is likely.

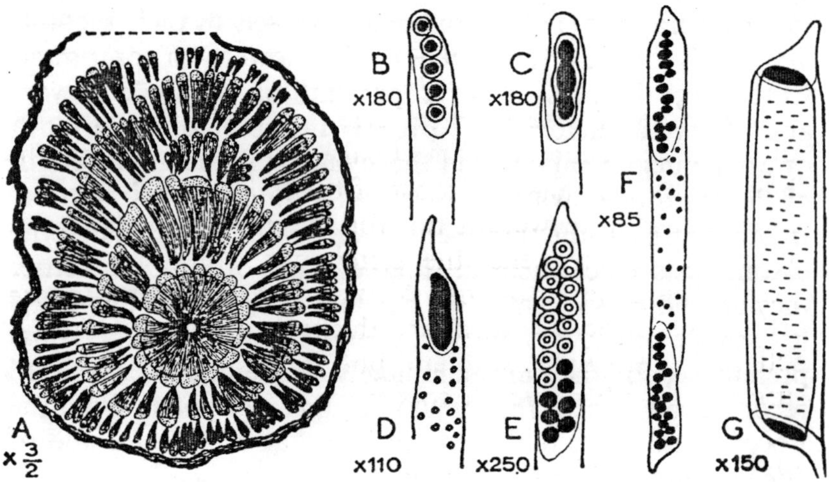

FIG. 37

Vascular tissue of Gnetales

A, t.s. stem of *Gnetum scandens*, showing anomalous secondary thickening. B–D, vessels of *G. africanum*. E, vessel of *Ephedra major*. F, vessel of *E. californica*. G, vessel of *Welwitschia Bainesii*. (B–D after Duthie; E, Thompson; F, Esau)

A further reason is that there are some primitive angiosperms that are completely without vessels—homoxylous—and, hence, it becomes necessary to seek an ancestor without vessels, instead of one with them.

Gnetaceae

There are up to forty species of *Gnetum*, most of which are lianes, although a few species, such as *G. gnemon*, are trees, and some are shrubs. The genus is restricted to the moist tropical forests of the Amazon basin, West Africa,

India, south China and Malaysia, *G. gnemon* being cultivated in Malaysia for its edible seeds. Of the many remarkable features shown by *Gnetum*, perhaps the most striking is the appearance of the leaves (Fig. 38A), which are indistinguishable from those of an angiosperm in having a broad lamina and reticulate venation. According to Takeda[171] and Florin,[62] the stomata are syndetocheilic, like those of *Welwitschia* (and those of the Bennettitales), but recently Maheshwari and Vasil[128] have claimed that, although apparently syndetocheilic, they are in fact haplocheilic.

The stem apex has a tunica layer (one cell thick) in which all the cell-divisions are anticlinal,[103] and a corpus, in the flanking regions of which most of the cell-divisions are also anticlinal. Indeed, were it not for the presence of a group of central mother-cells, the stem apex would be almost indistinguishable from that of many angiosperms. In the arborescent species, secondary thickening results from the activities of a single cambium, but, in the climbing species, successive cambia give rise to a series of co-axial cylinders of xylem and phloem (Fig. 37A), as in many angiospermous lianes.

In some species the secondary phloem (unlike that of most gymnosperms) contains companion cells, as well as sieve-tubes, but these two types of cell arise from different cambial initials, and not from the subdivision of sieve-tube mother-cells (as in angiosperms).[183]

The reproductive organs, which for want of a better term are usually called 'flowers', are borne in whorls on spike-like inflorescence axes, each whorl apparently subtended by a fleshy collar (Figs. 38C and 39A). Some species (e.g. *G. africanum*) are invariably dioecious, but in others (e.g. *G. gnemon*) the male inflorescences bear ovules which are normally sterile, but which may occasionally be fertile. Examination of young inflorescences[197] (Figs 38B and 39B) shows that the flower-primordia arise, not, as was thought for a long time, in the axils of the successive collars, but on their abaxial sides.

Each female flower consists of a nucellus, surrounded by three envelopes, described by some as three integuments and by others as two integuments and a perianth. The three layers arise in acropetal sequence (Figs. 38E–H), but the innermost eventually extends beyond the others as a micropylar tube. It is fused to the lower half of the nucellus and in some species develops a flange over the top of the middle envelope. In the mature seed the middle envelope is stony and the outermost is fleshy. All three envelopes receive a vascular supply.

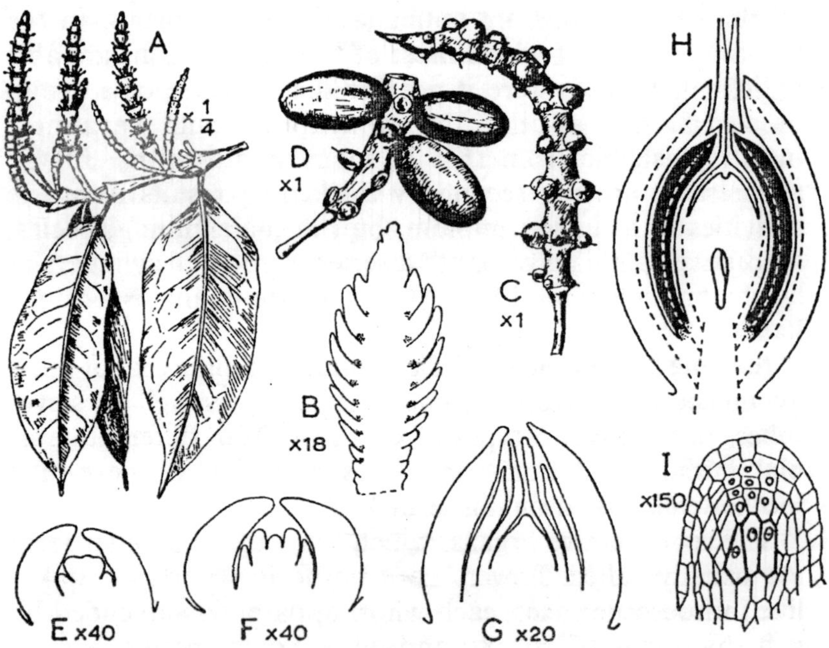

FIG. 38

Gnetum

Gnetum gnemon: A, leafy shoot with male inflorescences; H, l.s. ovule (diagrammatic); I, l.s. nucellus, showing three megaspore-mother-cells. *Gnetum africanum:* B–D, stages in development of the female inflorescence; E–G, stages in development of the ovule. (A, after Firbas, modified from Karsten and Liebisch; B–G, Waterkeyn; I, Strasburger; H, based on Pearson)

At the stage of initiation of the inner envelope, several archesporial cells appear, whose subsequent periclinal divisions give rows of parietal cells and megaspore-mother-cells (Fig. 38I). At one time these divisions were assumed to be meiotic and to give rise to linear tetrads of spores. It is now known, however, that the female gametophytes are tetrasporic in origin,[197] a condition which is shared with *Welwitschia* and with several angiosperm genera, but which is not known elsewhere among gymnosperms. Within each megaspore-mother-cell (of which there may be as many as eight to ten) meiosis proceeds without the laying down of cross-walls. Only two or three develop beyond this point, the nuclei undergoing numerous mitotic divisions, but still without cross-wall formation. Ultimately, the chalazal end of each prothallus becomes cellular (Fig. 39G), but the micropylar end retains the free-nuclear condition until after fertilisation has taken place.

Stages in the development of the male inflorescences are illustrated in Figs. 39B, C and A. Whereas there is only one whorl of flowers between adjacent collars on the female inflorescences, in the male there are several whorls of flowers which arise in basipetal succession (Fig. 39C). Each male flower consists of a single microsporangiophore (bearing one or two sporangia) surrounded by a tubular perianth (Fig. 39F). In those species that possess them, the abortive ovules in the male inflorescences usually have only two envelopes round the sterile nucellus (Fig. 39E).

At pollination time the pollen-grains contain only three nuclei, whose morphological nature has been the subject of much discussion. There has even been some doubt as to whether any of them is surrounded by its own cell-wall. One current view[164] is that prothallial cells are lacking, and that the three nuclei represent the tube-nucleus, a sterile cell and a spermatogenous cell (the latter destined to give rise to two non-motile male cells).

By means of a drop-mechanism the pollen is drawn down the micropyle, whose lower end then becomes plugged by an 'obturator'.[197] No archegonia are formed, and which of the prothallial nuclei are destined to act as egg-nuclei cannot be discerned. It is customary to say that any one of them can become an egg-nucleus in the event of a pollen-tube touching it.

Several pollen-tubes may penetrate each prothallus, resulting in the formation of many zygotes, each of which then divides to form two primary suspensor tubes. These then elongate considerably, branching as they do so, until a very large number of such tubes is formed. From the tip of

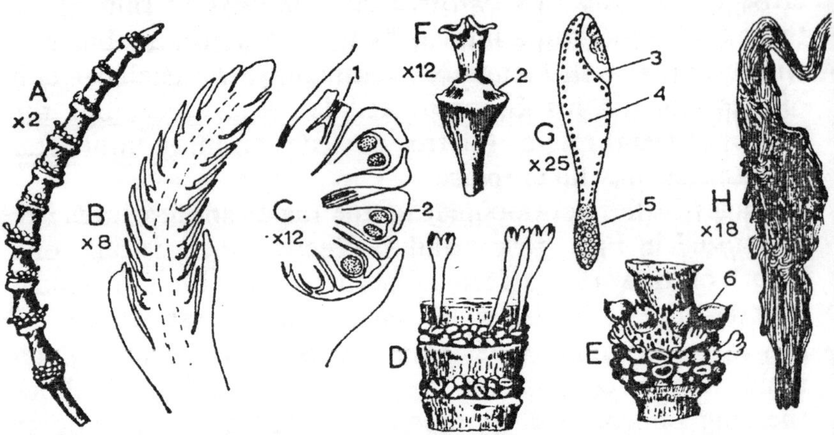

FIG. 39
Gnetum (continued)

Gnetum africanum: A, young male inflorescence; B, early stages in development of the male inflorescence; F, male flower with dehisced pollen-sacs, and perianth (2). *G. gnemon:* C, l.s. node and internode of a young male inflorescence, showing four male flowers at different stages of development, and one young abortive ovule (1); E, part of male inflorescence, showing abortive ovules (6); G, two female prothalli, one abortive (3), and one functional with nuclear (4) and cellular (5) regions. *G. scandens:* D, part of a male inflorescence with exserted stamens; H, secondary suspensor, exhibiting abundant polyembryony (embryos in black).

(A, B, F, after Waterkeyn; C, Coulter and Chamberlain; D. E. Griffith; G, Lotsy; H, Vasil)

each is produced a multicellular secondary suspensor, at the end of which is the embryo proper. Polyembryony is, therefore, of a very high order in *Gnetum*—not only are there several prothalli in each seed and several zygotes in each, there is also the multiplication of embryos from each zygote by branching of the primary suspensors, and further proliferation of the secondary suspensors can occur, as illustrated in Fig. 39H; yet only one embryo normally reaches maturity in each seed.

The embryo is peculiar in possessing a lateral finger-like process, called a 'feeder', which remains embedded in the seed, long after germination, and which, presumably, has an absorptive function. A similar structure is present in *Welwitschia*, but is absent from *Ephedra*, as, indeed, it is from all other gymnosperms.

Welwitschiaceae

This family is represented by the single species *Welwitschia Bainesii* (= *mirabilis*), which is restricted in its distribution to a narrow coastal belt about 600 miles long in south-west Africa.[148] Chamberlain[5] called it 'the most bizarre of all gymnosperms if not of all seed-plants', and with justification. The plant resembles a gigantic turnip, reaching a diameter of more than a metre, with two opposite leaves that grow continuously from a basal meristem throughout the life of the plant.

The apical meristem of the young seedling is minute and, because of frequent periclinal cell-divisions at all levels, there is no clearly marked zonation.[148] But, in any case, the apical meristem soon ceases to function and dies, further growth being restricted to the periphery, resulting merely in increased girth. As a result of recent work by Martens and Waterkeyn,[132] it is now known that three pairs of foliar organs arise, in an opposite and decussate sequence, instead of two as was hitherto believed. There are two cotyledons, which persist for a short time. Then come the two principal leaves and, just before the apical meristem aborts, two further leaf-primordia appear. These give rise to the so-called 'scaly bodies', which have been interpreted variously, by earlier workers, as cones, or buds, in the axils of the cotyledons.

The specimen illustrated in Fig. 40A is a relatively young one. In older plants the leaves become split into longitudinal ribbons, and the tips become worn away, so that it is rare to find leaves longer than 2 m, even in plants more than 100 years old. Their main venation is parallel, as would be expected in view of their mode of growth, but there are in addition small anastomosing lateral veins,[148] and there are large numbers of lignified spicular cells, encrusted with crystals of calcium oxalate. The stomata are syndetocheilic, as in Bennettitales.[62]

The upper surface of the stem becomes covered with a thick, ridged corky layer. In young stems there is a ring of vascular bundles, but in older ones there is a saucer-shaped mass, from which vascular traces go out to the leaves, the inflorescences and to the single tap-root. In older plants still, there are successive zones of xylem and phloem, as in *Cycas* and some species of *Gnetum*.

Welwitschia is invariably dioecious, and the inflorescences arise from a series of transverse ridges, parallel to the leaf-bases. They branch in a dichasial manner, each branch terminating in a beautifully regular cone, with opposite and decussate cone-scales (or bracts?) (Figs. 40B and 41A), which are a vivid crimson or scarlet colour at maturity. Ever since Hooker[101] first described the plant in 1862 there have been discussions about the correct interpretation of the flowers, but the most recent and exhaustive investigation of their ontogeny, by Martens,[131] must surely have settled them. This elegant work shows that the male and female flowers are both built up on the same opposite and decussate plan. This is emphasised by a comparison of the two floral diagrams (Figs. 40C and 41C).

Each female flower (Figs. 41B and D) is subtended by a cone-scale, and consists of a single nucellus, enclosed by two envelopes, and with two small lateral bract-like structures. Of the two envelopes, the inner is prolonged into a long tubular micropyle, and is undoubtedly a true integument. The outer is also called an integument by some morphologists, but it originates from two posterior-anterior primordia, that soon become confluent. By some, therefore,

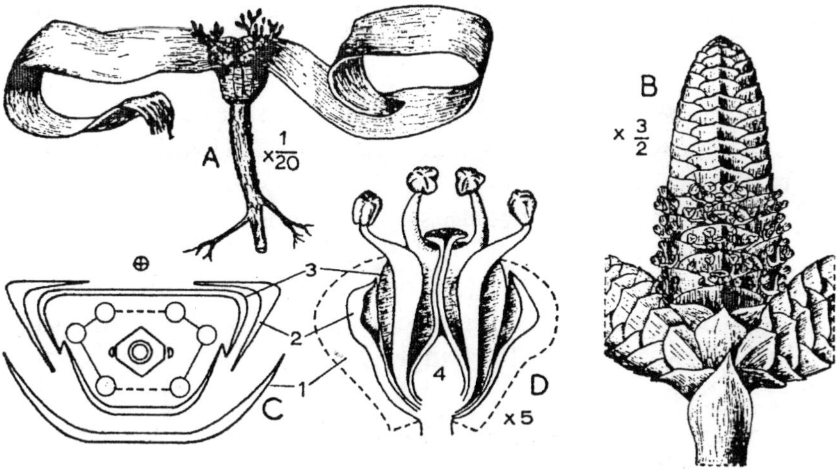

FIG. 40
Welwitschia Bainesii

A, young plant with cones. B, male cone. C, floral diagram of male flower. D, half-flower (male).

(1, subtending bract; 2, lateral bracts; 3, perianth; 4, nucellus of sterile ovule)

(A, after Eichler; B, D, Church; C, Martens)

it is called a perianth. In the mature seed it is expanded into a broad wing, which plays a part in wind-dispersal.

A single megaspore-mother-cell is formed low down in the nucellus, but, instead of giving a linear tetrad of spores, it develops directly into a tetrasporic female prothallus, for no cross-walls are laid down during meiosis.[131] Free-nuclear mitotic divisions then take place until well over 1,000 nuclei have been formed, before cell-walls are laid down. This takes place rather irregularly, with the result that the cells contain varying numbers of nuclei, which are said to fuse subsequently, to give polyploids of varying degrees. No archegonia are formed.

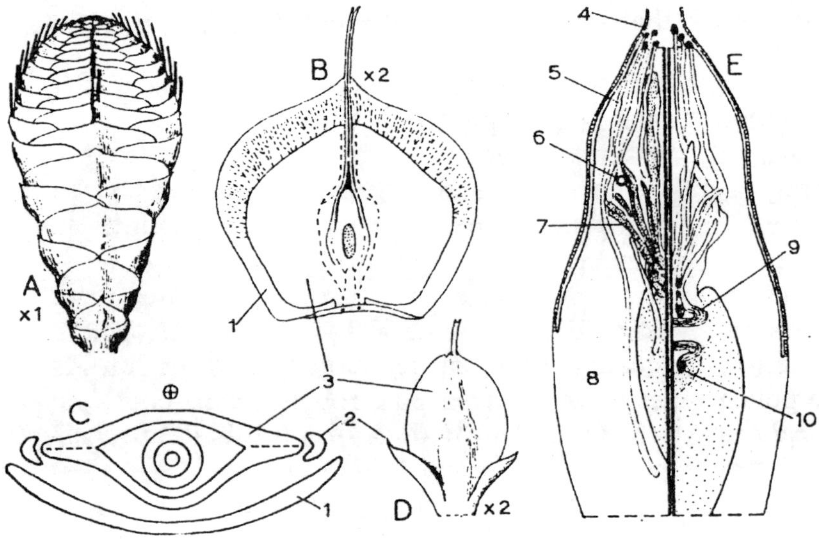

FIG. 41

Welwitschia Bainesii (continued)

A, ovuliferous cone. B, mature female flower. C, floral diagram of female flower. D, young female flower. E, semi-diagrammatic representation of ovule, at two stages of fertilisation (left-hand side at the stage of fertilisation; right-hand side with embryos). (1, subtending bract; 2, lateral bracts; 3, perianth; 4, pollen-grains; 5, pollen-tubes; 6, zygote; 7, prothallial tubes; 8, nucellus; 9, suspensor; 10, embryo. Prothallus stippled)

(A, B, after Church; C, D, Martens; E, Pearson)

Each male flower is subtended by a cone-scale. There are two lateral bracts (Fig. 40D) and a perianth, formed from two connate posterior-anterior bract-like structures. Then comes a whorl of six microsporangiophores, fused at the base into a cup-like structure, strongly reminiscent of the Bennettitales, and in the centre is a sterile ovule with a single integument. Each microsporangiophore bears a synangium of three fused sporangia, from which the pollen-grains are shed through a vertical slit. A honey-like fluid is secreted by the flared micropyle, and it is thought that insects may be attracted by it, although whether pollination is effected by insects or by wind appears not to have been established.

Little is known of the sequence of events leading to the three-celled male gametophyte,[164] but, by analogy with *Gnetum*, it is assumed that there are no prothallial cells—merely a tube-nucleus, a sterile cell (which aborts, even before pollination) and a spermatogenous cell, destined to give two sperm nuclei.

The fertilisation process in *Welwitschia* is unique. Two stages are illustrated diagrammatically in Fig. 41E, where the left-hand side shows an earlier stage than the right. As the pollen-tubes grow downwards through the nucellus, the apical cells of the female prothallus form 'prothallial tubes' which grow upwards to meet them. Contact is made, somewhere in the nucellar cap, and the female nuclei are said to pass into the pollen-tubes, where nuclear fusion takes place. What degree of polyploidy occurs in the zygote is not known, nor is it understood how the number of chromosomes becomes reduced. The zygote-nucleus divides to give a two-celled pro-embryo, of which the upper cell becomes a suspensor, and the lower divides further to give a number of secondary suspensor cells and a multicellular embryo, which is protected by a cap of eight cells as it is pushed down into the prothallus. As many zygotes are produced, there is a high degree of polyembryony, but only one embryo develops to maturity in any one seed.

Ephedraceae

The genus *Ephedra* is represented by some forty species, most of which are shrubby 'switch-plants', although one species may grow into a small tree, and a few are lianes. Many spread by means of rhizomes that grow from underground buds.[9] Representatives occur in North and in South America and in the Old World, where they extend in a broad belt from the Mediterranean to China. Several Asiatic species are important medicinally and have been used in China for over 5,000 years as a source of the drug ephedrin.

In most species the leaves are opposite and decussate, but in some they are in whorls of three or four. Accordingly, the number of primary vascular strands in the stem varies from species to species. Most commonly, the number in each internode is eight (two pairs of 'foliar traces' which run up to the node above before passing out into the leaves, and four 'stem-bundles'). Each leaf receives two bundles which have quite distinct points of origin in the stem system[130] (an arrangement described as 'two-trace, unilacunar', as in *Ginkgo* and in some pteridosperms). The leaves are reduced to minute scales which are soon shed, by means of an absciss layer, and photosynthesis is carried out by the green, ribbed stems.

The apical meristem is like that of *Gnetum* in having a well-marked tunica layer, but the stem is also capable of elongation by means of an intercalary meristem at the base of each internode. A single persistent cambium gives rise to secondary xylem, which is markedly 'ring-porous', and to secondary phloem, which is like that of *Gnetum* in that sieve-tubes and companion-cells are formed in rows, each from different cambial initials.

Ephedra is typically dioecious, but there are occasional reports of bisexual inflorescences. Even hermaphrodite flowers have been reported, but Eames[58] regards these as monstrosities.

The female inflorescence consists of a short shoot, bearing two to four pairs (or whorls) of bracts, which in some species become swollen and juicy (Fig. 42C). At the apex of the short shoot there are one to three female 'flowers', each consisting of a nucellus enclosed by two envelopes (interpreted either as two integuments or as one integument and a perianth). An early stage in development is illustrated in Fig. 42D, in which the apex of the short shoot can be discerned between the two young flowers. At maturity the lower half of the inner envelope is fused to the nucellus, but the upper half is free, and is prolonged into a long micropyle. This envelope receives two vascular bundles which, in some species, extend to the level of separation from the nucellus, but which in others die out near the base (Fig. 42F). Eames concludes that there has been a phylogenetic fusion of two organs in the evolution of this integument. Likewise, he concludes that the outer envelope represents an anterior-posterior pair of bracteoles that have become fused. At maturity it becomes hard and husk-like. Commonly it receives two or three vascular bundles. In species with only a single female flower, however, there may be four or six, and it is generally believed that the extra bundles belong to an abortive second flower.

The development of the young ovule is much less extraordinary than in *Gnetum* and *Welwitschia*, being much more like that in a typical gymnosperm. A linear tetrad of spores is produced from a single megaspore-mother-cell, and it is the lowermost that enlarges, to produce the female gametophyte. Free-nuclear divisions occur, until 256 or 512 nuclei have been formed, before wall-formation occurs, giving a cellular prothallus. Two or three archegonia are formed (or occasionally only one), each arising from a superficial cell, which divides periclinally into a primary neck-cell and a central cell (Fig. 42E). The apex of the female prothallus grows up towards the micropyle, forming a papilla, and the archegonial necks keep pace, becoming as much as forty cells long[127] (Fig. 42G). The central cell enlarges and its nucleus divides to give a ventral-canal-nucleus and an egg-nucleus.

The male flowers are borne in compound inflorescences (Fig. 42A). Each flower is subtended by a bract and has a two-lipped perianth (Fig. 42B). In many species there is a single central microsporangiophore (often called a stamen, or an antherophore) bearing a group of septate sporangia. In others the microsporangiophore is forked,[172] while in *E. distachya* and *E. intermedia* there are two microsporangiophores. Eames[58] regards the latter as the primary condition, and believes that the other species show varying degrees of fusion. The microsporangia have either two or three loculi, each opening by a terminal slit to release the pollen-grains.

The stages in the formation of the male prothallus are similar to those in *Pinus*, giving two prothallial cells, a tube-

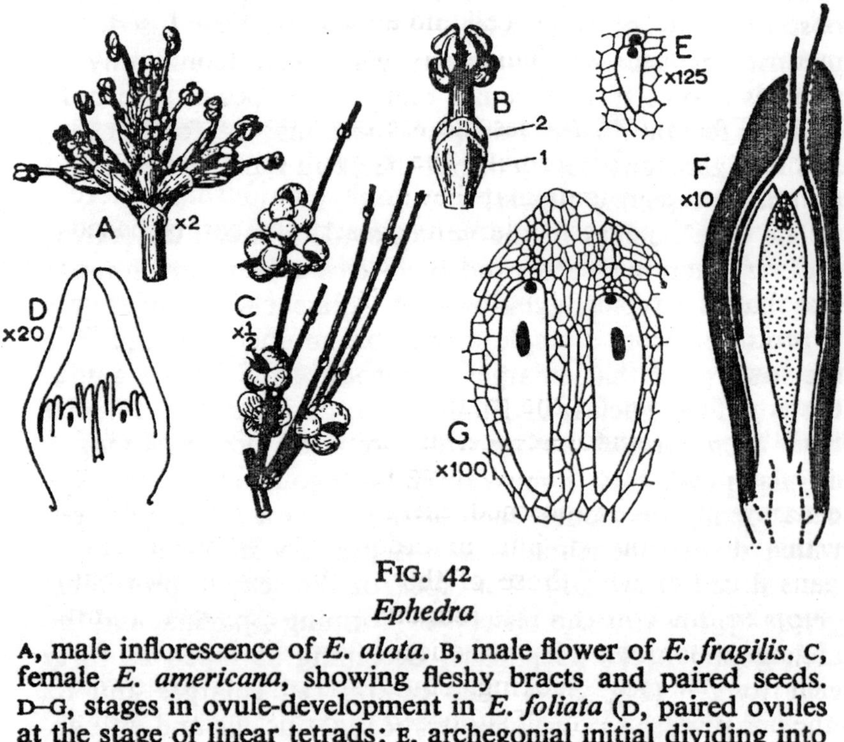

FIG. 42
Ephedra

A, male inflorescence of *E. alata*. B, male flower of *E. fragilis*. C, female *E. americana*, showing fleshy bracts and paired seeds. D–G, stages in ovule-development in *E. foliata* (D, paired ovules at the stage of linear tetrads; E, archegonial initial dividing into primary neck-cell and central cell; F, mature ovule with large pollen-chamber; G, apex of prothallus with mature archegonia)

(1, bract; 2, perianth)

(A, B, after Eichler; C, Le Maout and Decaisne; D, E, G, Maheshwari; F, based on Maheshwari)

nucleus, a sterile cell and a spermatogenous cell. Pollination involves a drop-mechanism, and the pollen-grains germinate within a few hours of entering the pollen-chamber. A pollen-tube penetrates the archegonium, and two nuclei are released, one of which fuses with the egg-nucleus. The other may fuse with the ventral-canal-nucleus ('double fertilisation'), but no embryo results from this second fusion.[107]

The zygote undergoes three divisions, the eight products of which are all capable of giving rise to embryos, although more commonly only three to five do so. Such precocious polyembryony is remarkable. Each pro-embryonal cell divides to give a suspensor-cell and an embryo-initial, which is pushed down into the female prothallus by the elongating suspensor.

Eames[58] strongly criticises the use of angiosperm-centred terminology when describing the reproductive organs of *Ephedra*, on the grounds that they imply relationships where none exists, and suggests alternative descriptive terms, based on his belief that this genus, unlike *Gnetum* or *Welwitschia*, has affinities with the Cordaitales. Thus, the male flower is interpreted as a fertile short shoot with, basically, two fertile appendages (sporophylls) and two sterile appendages, in the axil of a bract on the main axis of a compound cone. Thus, the whole inflorescence, as shown in Fig. 42A, consists of a 'cone-cluster', containing three cones. In so far as the male organs are concerned, there is certainly justification for this interpretation, but, in order to bring the female organs into line with those of the Cordaitales, Eames has to suppose that considerable reduction has taken place, and he figures a number of hypothetical intermediate stages, for which real evidence is lacking. Thus, each of the two flowers in the inflorescence is itself supposed to be all that is left of a fertile short shoot, with sterile appendages and two fertile ones (sporophylls bearing ovules). One of the fertile appendages has been suppressed, along with the apex of the short shoot, so that the remaining fertile appendage takes up a terminal position, and the sterile appendages become the outer envelope, or perianth. Whatever the truth of this, it is clear that *Ephedra* is morphologically very unlike *Gnetum* and *Welwitschia*, and may well have had a completely different origin.

Haploid chromosome numbers of 7 and 14 are characteristic of *Ephedra*, whereas in *Welwitschia* the number is 21, and in *Gnetum* (only two species of which have been investigated) it is 22.[108]

12

General Conclusions

In attempting to draw general conclusions about such a heterogeneous group as the gymnosperms the most difficult problem is to decide how far it is justifiable to go beyond the facts into the realms of speculation. From the previous chapters it is possible to detect several overall evolutionary trends which have been operating from the Lower Carboniferous to the present day. One is, however, tempted to go further than this and to extrapolate these trends backwards, with a view to discovering the kind of plants from which the gymnosperms might have evolved. Some may criticise a procedure of this kind as unscientific. Others, however, may be prepared to take it for what it is worth—the putting forward of a hypothesis which, one day, may be tested against the facts, by the discovery of more fossils, but which, in the meantime, is no more than a hypothesis, to be accepted or rejected by the reader as he pleases.

The main evolutionary trend in stem anatomy has been the suppression of the primary wood, leading to a system of discrete xylem strands associated with the origin of leaf-traces—i.e. a change from cauline to foliar. Indeed, the earliest members of the group showed much less distinction between leaves and stems, a fact that is in keeping with the Telome Theory. Lam[118] has attached much importance to the way in which the reproductive organs are borne, and recognises a distinction between the 'phyllosporous' and the 'stachyosporous' arrangement. However, the distinction is only one of degree and is not a fundamental one. Originally all reproductive organs were, presumably, stachyosporous; some have remained so, while others have become phyllosporous as the structures bearing them have become progressively more leaf-like.

At the same time, there has been a marked trend in the direction of greater protection of the megasporangium, involving the evolution of an integument (whatever its morphological nature may be), a cupule and yet further envelopes, such as the aril, an outer integument, perianth, etc. In addition, cones or cone-like structures have evolved, and further protection has been provided in some groups by the inversion of the ovules and the sterilisation of potentially fertile regions. *Ginkgo* is a notable exception, with its ovules fully exposed, but it may be argued that this genus is relatively primitive in several other respects, even though it is alive today. Such exceptions do not destroy the validity of the generalisation about progressive protection.

Extrapolating backwards, one would conclude that the earliest gymnosperms probably had stems with solid primary wood, presumably with some secondary wood also, and that there was little distinction between leaf and stem, the seeds and pollen-bearing organs being borne, fully exposed, at the tips of photosynthetic branchlets.

There still remains, however, the question as to whether seeds evolved more than once, bilateral ones being associated with pycnoxylic wood, and radial ones with manoxylic wood. Until very recently, opinion had hardened in favour of a diphyletic (or even polyphyletic) origin. However, several fossil genera have already been mentioned, in Chapter 7, which lead one to doubt this conclusion, e.g. Poroxylaceae and Eristophytaceae, and there is the recent suggestion by Smith[163] that even the bilateral cordaitalean seed might have evolved from one with radial symmetry.

One's doubts are further strengthened by a study of a heterogeneous assemblage of early fossils which show certain anatomical characters commonly associated with gymnosperms, but which, for various reasons, are thought to have fallen short of full gymnosperm status. For convenience, these have been grouped together as the Progymnospermopsida, and are placed in three orders, as follows:

1. **Pityales***	*Archaeopteris* (+ *Callixylon*) *Pitys* (?) *Archaeopitys* (?)
2. **Protopityales***	*Protopitys*
3. **Aneurophytales***	*Aneurophyton* (*Eospermatopteris*?), *Textraxylopteris* (= *Sphenoxylon*?) *Rellimia* (= *Milleria*, = *Protopteridium*), *Triloboxylon, Proteokalon,*

This group was first proposed by Beck,[42] when organic connection was first conclusively demonstrated between the stem genus *Callixylon* and the 'frond' genus *Archaeopteris*, one of the most important discoveries of recent years. Some five species of *Callixylon* are known, of Devonian and Lower Carboniferous age, most of them being found in North America, and one in Russia. Trunks up to 1·5 m in diameter have been found and it is estimated that their overall height may have approached 20 m. The bulk of the trunk was made up of secondary wood whose tracheids were pitted in a most peculiar manner. The pits on the radial walls were in groups separated by unpitted regions, and the pattern was the same on adjacent tracheids, so that, in longitudinal radial sections, one sees continuous horizontal bands of pits separated by unpitted bands. In most species the wood-rays were only one (or sometimes two) cell wide, and the wood was, therefore, typically pycnoxylic, but in one species (*C. Newberryi*) the rays were as much as four cells wide. In the centre of the trunk was a pith region 1 cm or more across, with a ring of numerous mesarch circum-medullary strands, mostly in contact with the secondary wood, but with some more deeply embedded in the pith. Each leaf-trace arose as a branch from one of these circum-medullary bundles. Except for the peculiar pitting of *Callixylon*, these trunks were so similar to those of the Coniferopsida that they were almost universally believed to have been early members of this group, despite the fact that no *bona fide* seeds had yet been found in the Devonian, and it was generally accepted that when its leaves were discovered they would be strap-shaped or needle-shaped. It came as a great shock, therefore, when its photosynthetic organs were shown to be large frond-like structures belonging to the genus *Archaeopteris*.

Some six species of *Archaeopteris* are known, all of which superficially resemble fern fronds, some being at least a metre long, branching bi-pinnately, with the pinnae apparently all in one plane. However, recent work[205, 207, 213] has shown that this interpretation is quite incorrect, and that *Archaeopteris* really consisted of a branch system bearing small spirally arranged leaves. On splitting a piece of rock containing a specimen of *Archaeopteris* one sees only those leaves that are in the plane of the fracture, with the result that they look like the pinnules of a frond. Those leaves which are not in the plane of the fracture can only be discovered by 'dégagement' of the rock, grain by grain.

Each leaf received a single vascular bundle which branched dichotomously, there being no midrib. In some species the sterile leaves were entire and fan-shaped, while in others they were deeply divided and laciniate. They were, indeed, quite similar to the leaves of the early fossil coniferophyte genera, *Buriadia* and *Carpenteria*, as Beck[205] has remarked. The fertile leaves were always finely divided and bore numerous elongated sporangia, in two rows, on the adaxial side. Dehiscence took place by means of a longitudinal slit.

At least three species of *Archaeopteris* are now known to have been heterosporous. This was demonstrated first in *A. latifolia* which had two kinds of sporangia, both about 2 mm long but differing in diameter. The narrow ones (0·3 mm in diameter) contained a hundred or more spores about 35 μ in size, while the broader ones (0·5 mm in diameter) contained up to sixteen spores about ten times larger. More recently[213] it has been established that *A. halliana* and *A. macilenta* also had megaspores of comparable sizes. They averaged 285 μ and 271 μ respectively, while their microspores averaged 42 μ. There can be no doubt that these species were heterosporous. Since microspores and megaspores were borne on one and the same individual, they cannot therefore have been seed-plants and must have been at the evolutionary level of pteridophytes. It is of particular interest that *A. macilenta* should be among the species proven to be heterosporous, for it was this species which Beck showed to be in organic connection with *Callixylon Zalesskyi*, whose massive development of secondary wood is therefore most surprising for a pteridophyte.

Callixylon has usually been classified along with the two Lower Carboniferous stem genera *Pitys* and *Archaeopitys*, and the question now arises as to whether these, too, are to be included in the Progymnospermopsida as pteridophytes. Unfortunately, no reproductive organs of any kind have been found in organic connection with them, nor even have leaves, with the one exception of *Pitys Dayi*. Like *Callixylon*, *Pitys* grew to large dimensions. *P. Withamii*, the 'Craigleith Tree', of which a trunk is set up in the grounds of the Natural History Museum in London, must have been much more than 20 m tall, because a 15 m specimen was found whose narrow end was as much as 45 cm across. At least five species are known, all of which had a relatively wide central pith, up to 5 cm across; in some species there was a mixed pith of parenchyma and tracheids. Both medullary and circum-medullary mesarch primary bundles were numerous, there being as many as fifty, even in a small stem. Each leaf-trace arose as a branch from one of these circum-medullary bundles and subdivided into three on its way to the leaf-base. The secondary wood varied considerably from species to species, for in some (e.g. *P. Withamii*) it was typically pycnoxylic, like that of a modern *Araucaria*, with wood-rays only two or three cells wide, whereas others approached the manoxylic condition (e.g. *P. primaeva* and *P. rotunda*) with rays up to fifteen cells wide.

In 1935 Gordon[76] found some leaves attached to young stems of *Pitys Dayi.* He described them as short fleshy structures, only 5 cm long and 5 mm across at the base. If these were complete leaves, with no parts missing, then not only were they quite unlike the 'fronds' of *Archaeopteris*, but they were also unlike the paddle-shaped leaves of the Cordaitales. However, as Beck has pointed out, their vascular bundles had a considerable quantity of secondary xylem, which suggests a megaphyllous type of plant. The problem has been made even more fascinating by a recent paper by Long,[124] who draws attention to the frequent occurrence together of *Pitys primaeva* stems and the petioles known as *Lyginorachis papilio.* Furthermore, he has shown how similar is the vascular system of these petioles to that of the 'leaf' of *Pitys Dayi.* Now, such petioles were part of large fern-like fronds which are usually assigned to the lyginopterid pteridosperms, and one of Long's specimens was shown to be continuous with yet another stem-like structure, *Tristichia Ovensii*, which formed a seed-bearing continuation of the rachis of the frond. This frond, therefore, clearly belonged, not to a pteridophyte, but to a gymnosperm. If it was indeed the frond of *Pitys*, then together they constituted a kind of plant unknown hitherto, for the pteridosperms did not have stout trunks. Either a new taxonomic group must be created, or the definition of the pteridosperms must be modified, so as to include such an extraordinary plant. *Archaeopitys*, represented by the one species *A. Eastmanii*, is so similar to *Pitys* that some palaeobotanists believe that it should be included in that genus.

Protopitys has been known for over a hundred years as a stem genus, but only recently has anything been known of its reproductive organs. Trunks of *P. Buchiana* are known to have attained a diameter of 45 cm, and must have grown to a considerable height. In the centre was an elliptical pith, surrounded by a narrow but continuous layer of metaxylem. Leaf-traces were given off alternately from the opposite ends of the ellipse. Presumably, therefore, the plant must have had a strange appearance, likened by some to the Travellers' Tree of Madagascar (*Ravenala madagascariensis*).[19] Nothing is known with certainty of the kind of leaf that *Protopitys* possessed, but it is very likely that it was large and frond-like. However, the secondary wood was pycnoxylic, with wood-rays that were only one or two cells wide and often only one cell high.

It has long been realised that *Protopitys* was an extremely isolated genus with a unique combination of vegetative characters. Its medullated protostele suggests an affinity with the ferns, its secondary wood with the Coniferopsida, its leaf-trace arrangement with the Cycadopsida. It is therefore of the greatest interest that some slight degree of heterospory has been suggested in *P. scotica*.[195] A small stem was discovered, bearing two lateral branch systems terminating in elongated sporangia about 3 mm long. These were relatively massive and appear to have possessed stomata in the sporangium wall. Most of the spores were 82 μ in diameter, but some of the sporangia contained larger ones, 147 μ across, while some bore spores of an intermediate size, 98 μ. Accordingly Walton suggested that *Protopitys* should be regarded as a pteridophyte exhibiting the early stages of evolution of heterospory.

The various genera now grouped together in the Aneurophytales had a lobed primary xylem, pycnoxylic secondary xylem in which the tracheids had pits on all the walls and three-dimensional lateral branch systems with terminal groups of sporangia. They extend from the lowest strata of the Middle Devonian, when *Rellimia* first appeared, to the middle of the Upper Devonian, when *Tetraxylopteris* was the last to become extinct. They exhibit several evolutionary trends that are paralleled in other groups, e.g. an increasing amount of parenchyma in the primary xylem and an increasing tendency for the ultimate parts of the branch system to become flattened and leaf-like.[214]

Rellimia[210, 211] when first described by Kräusel and Weyland[209] under the name *Protopteridium* was said to have had fern-like fronds, but these are now known to have been branch systems, arranged in a spiral instead of in one plane. Sterile parts bore 'leaves' up to 7 mm long, which bifurcated three times and had flattened tips. The fertile appendages were highly complex, for they forked into two and each half was then thrice-pinnate, with numerous sporangia at the tips of the pinnae.

It is now generally accepted that the two generic names *Aneurophyton* and *Eospermatopteris* are synonymous, the former being used by Kräusel and Weyland[115] for material collected from Upper Devonian rocks of Germany and the latter by Goldring[74] a year later for material of similar age from New York. *Aneurophyton* is the name which takes precedence, therefore. The American plants had trunks up to 13 m high, with a bulbous base, up to 1 m in diameter. The sterile appendages were minute leaf-like structures, forking once or twice, with recurved tips. The fertile ones were similar to those of *Rellimia*, but smaller and less complex. The sterile appendages of *Triloboxylon*[214] were much divided (four or more times) in one plane.

Tetraxylopteris[206] and *Proteokalon*[215] differed from earlier members of the group in that the lateral branch systems were based on an opposite and decussate plan. The sterile appendages of *Tetraxylopteris* were once or twice bilobed, while the reproductive organs dichotomized twice, each resulting branch being thrice-pinnate. The sterile appendages of *Proteokalon* showed a much greater degree of planation for they forked at least five times in one plane and their tips were flattened. Unfortunately the reproductive organs of *Proteokalon* are not known.

The fact that the photosynthetic branch systems of *Archaeopteris*, in the Pityales, and of both *Rellimia* and *Aneurophyton*, in the Aneurophytales were at first thought to be megaphyllous fronds leads one to speculate as to whether such branch systems might have been the evolutionary fore-runners of such fronds. If this were so, then the Progymnospermopsida are clearly of the greatest possible interest to the student of gymnosperm evolution, for they could be seen as combining some of the characters of the Cycadopsida with some of those of the Coniferosida; and the possibility may therefore be entertained that the gymnosperms are monophyletic. Confirmation must, however, await the discovery of further fossils from Devonian, or even Silurian deposits. The reader may feel sympathy with the Chinese philosopher who said, 'He who considers everything decides nothing', but surely this is the way of all scientific progress. As new facts come to light, theories must be altered to accomodate them. Morphology would indeed be as dead as the Dodo if theories of evolution were not constantly changing.

Bibliography

REFERENCE BOOKS

1 Andrews, H. N., 1961. *Studies in paleobotany*. New York and London.

2 Arber, A., 1950. *The natural philosophy of plant form*. Cambridge.

3 Arnold, C. A., 1947. *An introduction to paleobotany*. New York and London.

4 Bowen, R. N. C., 1958. *The exploration of time*. London.

5 Chamberlain, C. J., 1934. *Gymnosperms—Structure and evolution*. Chicago. (Reprinted, 1957, New York)

6 Coulter, J. M. and Chamberlain, C. J., 1917. *Morphology of gymnosperms*. Chicago.

7 Dallimore, W. and Jackson, A. B., 1948. *A handbook of Coniferae, including Ginkgoaceae*. London.

8 Delevoryas, T., 1962. *Morphology and evolution of fossil plants*. New York and London.

9 Foster, A. S. and Gifford, E. M., 1959. *Comparative morphology of vascular plants*. San Francisco.

10 Gaussen, H., 1950–2. *Les gymnospermes actuelles et fossiles*. Toulouse.

11 Greguss, P., 1955. *Identification of living gymnosperms on the basis of xylotomy*. Budapest.

12 Hort, A., 1949. *Theophrastus: Enquiry into plants, with an English translation*. London.

13 Jeffrey, E. C., 1917. *The anatomy of woody plants*. Chicago.

14 McClean, R. C. and Ivimey-Cook, W. R., 1951. *Textbook of theoretical botany*. Vol. 1. London.

15 Melchior, H. and Werdermann, E., 1954. A. Engler's *Syllabus der Pflanzenfamilien*. Vol. 1, *Bacterien-Gymnospermen*. Berlin.

16 Pearson, H. H. W., 1929. *Gnetales*. Cambridge.

17 Renault, B., 1881–5. *Cours de botanique fossile fait au Muséum d'Histoire naturelle*. 4 vols. Paris.

18 Schuster, J., 1932, in *Das Pflanzenreich*. Leipzig.

19 Scott, D. H., 1923. *Studies in fossil botany*. Vol. 2. *Spermophyta*. London.

20 Seward, A. C., 1898–1919. *Fossil plants, for students of botany and geology*. 4 vols. Cambridge.

21 Sprecher, A., 1907. *Le Ginkgo biloba*. Thesis, Geneva.

22 Thibout, E., 1896. *Recherches sur l'appareil mâle des gymnospermes*. Lille.

23 Walton, J., 1953. *An introduction to the study of fossil plants*. London.

24 Wieland, G. R., 1906 and 1916. *American fossil cycads*. 2 vols. Washington.

25 Zimmermann, W., 1959. *Die Phylogenie der Pflanzen*. Stuttgart.

PAPERS

26 Abraham, A. and Mathew, P. M., 1962. *Ann. Bot.*, *26*, 261–6. (X and Y chromosomes in *Cycas*)

26a Allen, G. S., 1946. *Am. J. Bot.*, *33*, 666–77. (Embryogeny in *Pseudotsuga*)

27 Andrews, H. N., 1940. *Ann. Mo. bot. Gdn.*, *27*, 51–118. (Wood-anatomy of pteridosperms)

28 Andrews, H. N., 1945. *Ann. Mo. bot. Gdn.*, *32*, 323–60. (*Schopfiastrum*)

29 Andrews, H. N., 1963. *Science*, *142*, 925–31. (Early seed-plants)

30 Andrews, H. N. and Felix, C. J., 1952. *Ann. Mo. bot. Gdn.*, *39*, 127–35. (Gametophyte of *Cardiocarpus*)

31 Antevs, E., 1914. *K. svenska VetenskAkad. Handl.*, *51*, 1–18. (Peltaspermaceae)

32 Arber, A., 1910. *Ann. Bot. 24*, 491–509. (*Mitrospermum*)

33 Arber, A., 1946. *Chronica bot.*, *10*, 67–123. (Goethe's botany)

34 Arber, E. A. N., 1905. *Quart. J. geol. Soc. Lond.*, *61*, 324–38. (Sporangia of *Glossopteris*)

35 Arber, E. A. N., and Parkin, J., 1907. *J. Linn. Soc.* (*Bot.*), *38*, 29–80. (Bennettitales and the origin of angiosperms)

35a Arnold, C. A., 1948. *Bot. Gaz.*, *110*, 2–12. (Classification of gymnosperms)

36 Arnold, C. A., 1953. *Phytomorph.*, *3*, 51–65. (Evolution of cycads)

37 Bailey, I. W., 1956. *J. Arnold Arbor.*, *37*, 269–87. (Nodal anatomy in retrospect)

38 Barnard, P. D. W., 1959. *Ann. Bot.*, *23*, 285–96. (*Eosperma*)

39 Barnard, P. D. W., 1960. *Palaeontology*, *3*, 265–75. (*Calathospermum fimbriatum*)

40 Baxter, R. W., 1949. *Ann. Mo. bot. Gdn.*, *36*, 287–352. (*Medullosa* spp.)

41 Beck, C. B., 1957. *Am. J. Bot.*, *44*, 350–67. (*Tetraxylopteris*)

42 Beck, C. B., 1960. *Brittonia*, *12*, 351–68. (*Callixylon* and *Archaeopteris*)

43 Benson, M. J., 1914. *Trans. roy. Soc. Edinb.*, *50*, 1–15. (*Sphaerostoma*)

44 Billings, W. D. and Thompson, J. H., 1957. *Ecology*, *38*, 158–60. (*Pinus aristata*, over 4,000 years old)

45 Brongniart, A., 1874. *Annls. Sci. nat.* (*Bot.*), *20*, 234–60. (Upper Carboniferous seeds)

46 Buchholtz, J. T. and Gray, N. E., 1948–51. *J. Arnold Arbor.*, *29*, 49–76, 117–51, *32*, 82–97. (*Podocarpus*)

47 Calder, M. G., 1938. *Trans. roy. Soc. Edinb.*, *59*, 309–31. (*Samaropsis*)

48 Camp. W. H. and Hubbard, M. M., 1963. *Am. J. Bot.*, *50*, 235–43. (Lyginopterid seeds)

49 Chowdhury, R., 1962, *Phytomorph. 12*, 313–38. (Conifer embryology)

50 Delevoryas, T., 1968 *Palaeontographica, 121*, 122–33. (Flowers of *Cycadeoidea*)

51 Delevoryas, T., 1956. *Contr. Mus. Geol. Univ. Mich., 12*, 285–99. (Shoot-apex of *Callistophyton*)

52 Delevoryas, T., 1959. *Am. J. Bot., 46*, 657–66. (*Cycadeoidea-Monanthesia*)

53 Delevoryas, T., 1963. *Am. J. Bot., 50*, 45–52. (Flowers of *Cycadeoidea*)

54 Delevoryas, T. and Morgan, J., 1954. *Palaeontographica, 96*, 12–23. (*Callistophyton*)

55 Doyle, J., 1945. *Sci. Proc. R. Dublin Soc., 24*, 43–62. (Pollination in conifers)

56 Doyle, J., 1963. *Proc. R. Irish Acad., 62 B*, 181–216. (Conifer embryology)

57 Dupler, A. W., 1920. *Bot. Gaz., 69*, 492–520. (*Taxus canadensis*)

58 Eames, A. J., 1952. *Phytomorph., 2*, 79–104. (Relationships of *Ephedra*)

59 Edwards, W. N., 1929. *Ann. Mag. nat. Hist., 4*, 385–94. (*Caytonia* in Sardinia)

60 De Ferré, Y., 1958. *Bull. Soc. bot. Fr., 105*, 155–205. (Extensive bibliography of conifer literature)

61 Florin, R., 1933. *K. svenska VetenskAkad. Handl., 12*, 1–134. (Mesozoic cycads)

62 Florin, R., 1934. *Svensk bot, Tidskr., 28*, 264–89. (Stomata of *Welwitschia*)

63 Florin, R., 1936. *Svensk bot. Tidskr., 30*, 624–51. (Pollen-grains of Cordaitales)

64 Florin, R., 1936–43. *Palaeontographica, 81*, 71–173, *82*, 1–72, *87*, 59–93. (Fossil Ginkgoales)

65 Florin, R., 1948. *Bot. Gaz., 110*, 31–9. (Taxaceae)

66 Florin, R., 1949. *Acta Horti Bergiani, 15*, 79–109. (*Tricho-pitys* and the evolution of Ginkgoales)

67 Florin, R., 1951. *Acta Horti Bergiani, 15*, 285–388. (Cordaitales and conifers)

68 Florin, R., 1955. In *A century of progress in the natural sciences, 1853–1953*, ed. E. L. Kessel. San Francisco. (Gymnosperm systematics, together with an extensive bibliography)

69 Florin, R., 1958. *Acta Horti Bergiani*, *17*, 257–402. (Jurassic taxads and conifers)

70 Florin, R., 1963. *Acta Horti Bergiani*, *20*, 121–312. (Distribution, past and present, of conifers and taxads)

71 Frenguelli, J., 1943. *Rev. Mus. La Plata*, *2*, 225–342. (Peltaspermaceae and Corystospermaceae from the Argentine)

72 Frenguelli, J., 1944. *An. Mus. La Plata*, *2*, 1–30. (Argentinian corystosperms)

73 Fry, W. L., 1955. *Bull. Torrey bot. Club*, *82*, 486–90. (*Cordaitanthus*)

74 Goldring, W., 1924. *Bull. N. Y. St. Mus.*, *251*, 50–72. (*Eospermatopteris*)

75 Gordon, W. T., 1912. *Trans. roy. Soc. Edinb.*, *48*, 813–25. (*Rhetinangium*)

76 Gordon, W. T., 1935. *Trans. roy. Soc. Edinb.*, *58*, 279–311. (*Pitys*)

77 Gordon, W. T., 1938. *Trans. roy. Soc. Edinb.*, *59*, 351–70. (*Tetrastichia*)

78 Gordon, W. T., 1941. *Trans. roy. Soc. Edinb.*, *60*, 427–61. (*Salpingostoma*)

79 Grand'Eury, F. C., 1877. *Mém. Acad. Sci., Paris*, *24*, 1–624. (Cordaitales)

80 Gray, N. E., 1953–62. *J. Arnold Arbor.*, *34*, 67–76, 163–75, *36*, 199–206, *37*, 160–72, *39*, 424–77, *41*, 36–9, *43*, 67–79. (*Podocarpus*)

81 Gunckel, J. E. and Wetmore, R. H., 1946. *Am. J. Bot.*, *33*, 532–43. (Nodal anatomy of *Ginkgo*)

82 Gunckel, J. E., Thimann, K. V. and Wetmore, R. H., 1949. *Am. J. Bot.*, *36*, 309–18. (Long and short shoots in *Ginkgo*)

83 Halket, A. C., 1930. *Ann. Bot.*, *44*, 865–905. (Rootlets of *Amyelon*)

84 Hall, J. W., 1954. *Bot. Gaz.*, *115*, 346–60. (*Stephanospermum* in American coal-balls)

85 Halle, T. G., 1929. *K. svenska VetenskAkad. Handl.*, *6*, 1–24. (Permian seed-plants of China)

86 Halle, T. G., 1935. *K. svenska VetenskAkad. Handl.*, *12*, 1–103. (Male organs of pteridosperms)

87 Halle, T. G., 1937. *C. r. Congr. Avanc. Etudes Stratigr. carb. Heerlen*, *1935*, 227–35. (Evolution of seeds and male organs in pteridosperms)

88 Harris, T. M., 1932. *Medd. Grønland*, *85*, 1–112. (Mesozoic pteridosperms and Bennettitales)

89 Harris, T. M., 1932. *Medd. Grønland* *85*, 1–133. (Caytoniaceae and Bennettitales)

90 Harris, T. M., 1933. *New Phytol.*, *32*, 97–114. (Caytoniaceae)

91 Harris, T. M., 1940. *Ann. Bot.*, *4*, 713–34. (*Caytonia*)

92 Harris, T. M., 1940. *Ann. Mag. nat. Hist.*, *6*, 249–65. (*Sagenopteris*)

93 Harris, T. M., 1941. *Ann. Bot.*, *5*, 47–58. (*Caytonanthus*)

94 Harris, T. M., 1941. *Phil. Trans.*, *B 231*, 75–98. (Fossil cycads)

95 Harris, T. M., 1943. *Ann. Mag. nat. Hist.*, *10*, 505–22. (*Anomozamites*)

96 Harris, T. M., 1944. *Phil. Trans.*, *B 231*, 313–28. (*Williamsoniella*)

97 Harris, T. M., 1951. *Phil. Trans.*, *B 235*, 483–508. (*Czekanowskia* and *Leptostrobus*)

98 Harris, T. M., 1951. *Phytomorph.*, *1*, 29–39. (Caytoniaceae)

99 Harris, T. M., 1961. *Palaeontology*, *4*, 313–23. (Fossil cycads)

99b Harris, T. M., 1962. *Trans. roy. Soc. N. Z. (Geol.)*, *1*, 17–27 (*Carnoconites* in New Zealand)

100 Holden, H. S., 1954. *Ann. Bot.*, *18*, 407–15. (*Physostoma stellatum*)

101 Hooker, J. D., 1862. *Trans. Linn. Soc. Lond.* (*Bot.*), *24*, 1–48. (*Welwitschia*)

102 Hoskins, J. H. and Cross, A. T., 1946. *Am. Midl. Nat.*, *36*, 207–50, 331–61. (*Pachytesta*)

103 Johnson, M. A., 1950. *Bull. Torrey bot. Club.*, *77*, 354–67. (Stem-apex of *Gnetum*)

104 Johnson, M. A., 1951. *Phytomorph*, *1*, 188–204. (Stem-apices of gymnosperms)

105 Kanis, A. and Karstens, W. K. H., 1963. *Acta bot. neerl.*, *12*, 281–86. (Amphistomatic leaves in *Ginkgo*)

106 Kern, E. M. and Andrews, H. N., 1946. *Ann. Mo. bot. Gdn.*, *33*, 291–306. (*Rhabdospermum and Kamaraspermum*)

107 Khan, R., 1943. *Proc. nat. Acad. Sci. India*, *13*, 357–75. (Fertilisation and embryology of *Ephedra foliata*)

108 Khoshoo, T. N., 1961. *Silvae Genet.*, *10*, 1–32. (Chromosome numbers in gymnosperms)

109 Khoshoo, T. N., 1962. *Proc. Summer Sch. Bot. Darjeeling*, 119–35. (Karyotypes of gymnosperms)

110 Kidston, R. and Gwynne-Vaughan, D. T., 1912. *Trans. roy. Soc. Edinb.*, *48*, 263–71. (*Stenomyelon*)

111 Kidston, R. and Lang, W. H., 1923. *Trans. roy. Soc. Edinb.*, *53*, 409–17. (*Palaeopitys*)

112 Kräusel, R., 1943. *Palaeontographica*, *B 87*, 59–93. (Fertile *Sphenobaiera*)

113 Kräusel, R., 1948. *Senckenbergiana*, *29*, 141–9. (*Sturiella*)

114 Kräusel, R., 1949. *Palaeontographica*, *B 89*, 35–82. (*Westersheimia* and *Williamsonia Wettsteinii*)

115 Kräusel, R. and Weyland, H., 1923. *Senckenbergiana*, *5*, 154–84. (*Aneurophyton*)

116 Kräusel, R. and Weyland, H., 1926 and 1929. *Abh. senckenb. naturforsch. Ges.*, *40*, 115–55, *41*, 315–60. (*Aneurophyton*)

117 Lacey, W. S., 1953. *Ann. Bot.*, *17*, 579–96. (*Eristophyton* and *Endoxylon*)

118 Lam, H. J., 1952. *Annls Biol.*, *28*, 57–88. (Classification of vascular plants)

119 Lee, C. L., 1954. *Am. J. Bot.*, *41*, 545–9. (Sex-chromosomes in *Ginkgo*)

120 Le Goc, M. J., 1914. *Ann. Bot.*, *28*, 183–93. (Centripetal xylem in cycad leaves)

121 Long, A. G., 1944. *Ann. Bot.*, *8*, 105–17. (Prothallus of *Lagenostoma*)

122 Long, A. G., 1960 and 1961. *Trans. roy. Soc. Edinb.*, *64*, 29–44, 201–15, 261–80, 281–95, 401–19. (Lower Carboniferous seeds)

123 Long, A. G., 1961. *Trans. roy. Soc. Edinb.*, *64*, 477–88. (*Tristichia*)

124 Long, A. G., 1963. *Trans. roy. Soc. Edinb.*, *65*, 211–24. (Fronds associated with *Pitys*)

125 Long, A. G., 1964. *Trans. roy. Soc. Edinb.*, *65*, 435–37. (*Stenomyelon* and *Kalymma*)

126 Lotsy, J., 1899. *Annls Jard. bot. Buitenz.*, *1*, 46–114. (Fertilisation in *Gnetum gnemon*)

127 Maheshwari, P., 1953. *Proc. Indian Acad. Sci.*, *1*, 586–606. (Gametophytes of *Ephedra*)

128 Maheshwari, P. and Vasil, V., 1961. *Ann. Bot.*, *25*, 313–19. (Stomata of *Gnetum*)

129 Mamay, S. H., 1954. *Prof. Pap. U. S. geol. Surv.*, *254–D*, 81–95. (*Tyliosperma*)

130 Marsden, M. P. F. and Steeves, T. A., 1955. *J. Arnold Arbor.*, *36*, 241–58. (Vascular system of *Ephedra*)

131 Martens, P., 1959, 1961 and 1963. *La Cellule*, *60*, 171–286, *62*, 7–91, *63*, 309–329. (*Welwitschia*)

132 Martens, P. and Waterkeyn, L., 1963. *Phytomorph.*, *13*, 359–63. (Stem-apex of *Welwitschia*)

133 Nakai, T., 1938. *Bull. For. Soc. Korea*, *19*, 1–29. (Quoted by Florin[70]) (Austrotaxaceae)

134 Nathorst, A. G., 1909. *K. svenska VetenskAkad. Handl.*, *45*, 1–37. (*Wielandiella* and *Williamsonia*)

135 Oliver, F. W., 1903. *Ann. Bot.*, *17*, 451–76. (Hyposperm in seeds)

136 Oliver, F. W., 1904. *Trans. Linn. Soc. Lond.* (*Bot.*), *6*, 361–400. (*Stephanospermum*)

137 Oliver, F. W., 1909. *Ann. Bot.*, *23*, 73–116. (*Physostoma*)

138 Oliver, F. W. and Salisbury, E. J., 1911. *Ann. Bot.*, *25*, 1–50. (*Conostoma* and *Gnetopsis*)

139 Oliver, F. W. and Scott, D. H., 1904. *Phil. Trans.*, *B 197*, 193–247. (*Lagenostoma*)

140 Pant, D. D., 1958. *Bull. Br. Mus. nat. Hist.* (*Geol.*), *3*, 127–75. (Glossopteridaceae)

141 Pant, D. D. and Mehra, B., 1964. *J. Linn. Soc.* (*Bot.*), *58*, 491–7. (Stomata in cycads and *Ginkgo*)

142 Pant, D. D. and Nautiyal, D. D., 1960. *Palaeontographica*, *B 107*, 41–64. (Seeds and pollen of Glossopteridaceae)

143 Phillips, T. L. and Andrews, H. N., 1963. *Ann. Mo. bot. Gdn.*, *50*, 29–51. (*Sutcliffia* in U.S.A.)

144 Plumstead, E. P., 1952, 1956 and 1958. *Trans. geol. Soc. S. Afr.*, *55*, 281–328, *59*, 211–36, *61*, 51–75, 81–94. (Glossopteridaceae)

145 Plumstead, E. P., 1956. *Palaeontographica*, *B 100*, 1–25. (Glossopteridaceae)

146 Read, C. B., 1936. *Prof. Pap. U.S. geol. Surv.*, *185–H*, 149–61. (*Diichnia*)

147 Read, C. B., 1936. *Prof. Pap. U.S. geol. Surv.*, *186–E*, 81–104. (*Stenomyelon* and *Calamopitys*)

148 Rodin, R. J., 1953 and 1958. *Am. J. Bot.*, *40*, 280–5, 371–8, *45*, 90–103. (*Welwitschia*)

148b Sahni, B., 1920. *Ann. Bot.*, *34*, 117–33. (Taxales)

149 Sahni, B., 1932. *Mem. geol. Surv. India*, *20*, 1–19. (*Williamsonia Sewardiana*)

150 Sahni, B., 1948. *Bot. Gaz.*, *110*, 47–80. (Pentoxylales)

151 Saxton, W. T., 1934. *Ann. Bot.*, *48*, 411–27. (*Austrotaxus*)

152 Schopf, J. M., 1949. *J. Paleontol.*, 22, 681–724. (*Dolerotheca*)

153 Scott, D. H., 1897. *Ann. Bot.*, *11*, 399–419. (Centripetal xylem in cycad cones)

154 Scott, D. H., 1906. *Trans. Linn. Soc. Lond.* (*Bot.*), *7*, 45–68. (*Sutcliffia*)

155 Scott, D. H. and Maslen, A. J., 1907. *Ann. Bot.*, *21*, 89–134. (*Trigonocarpus*)

156 Sen, J., 1955. *Bot. Notiser*, *108*, 244–52. (Fertile Glossopteridaceae)

157 Sen, J., 1958. *Bot. Notiser*, *111*, 436–48. (*Vertebraria*)

158 Seward, A. C., 1903. *Quart. J. geol. Soc. Lond.*, *59*, 217–33. (*Dictyozamites*)

159 Seward, A. C., 1938. *Sci. Progr.*, *127*, 420–40. (*Ginkgo*)

160 Shapiro, S., 1951. *Am. J. Bot.*, *38*, 47–53. (Stomata on the nucellus of *Zamia*)

161 Shimamura, T., 1937. *Int. J. Cytol.* (*Tokyo*), 416–23. (Sperms of *Ginkgo*)

162 Sitholey, R. V. and Bose, M. N., 1953. *Palaeobotanist*, *2*, 29–39. (*Williamsonia santalensis*)

163 Smith, D. L., 1964. *Biol. Rev.*, *39*, 137–59. (Ovule evolution)

164 Sterling, C., 1963. *Biol. Rev.*, *38*, 167–203. (Male gametophytes of gymnosperms)

165 Stewart, W. N., 1951. *Am. Midl. Nat.*, *46*, 717–42. (*Pachytesta hexangulata*)

166 Stewart, W.N., 1954. *Am. J. Bot.*, 41, 500-8. (*Pachytesta illinoense*)

167 Stewart, W. N., 1958. *Am. J. Bot.*, *45*, 580–8. (*Pachytesta composita*)

168 Stewart, W. N. and Delevoryas, T., 1956. *Bot. Rev.*, *22*, 45–80. (Medullosaceae)

169 Stopes, M. C., 1905. *Ann. Bot.*, *19*, 561–6. (Double nature of cycad integuments)

170 Surange, K. R. and Srivastava, P. N., 1956. *Palaeobotanist*, *5*, 46–9. (Glossopteridaceae)

171 Takeda, H., 1913. *Ann. Bot.*, *27*, 347–57, 365–6, 547–52. (Stomata in *Gnetum* and *Welwitschia*)

172 Thoday, M. G. and Berridge, E. M., 1912. *Ann. Bot.*, *26*, 953–85. (*Ephedra*, flowers and inflorescence)

173 Thomas, H. H., 1913. *Quart. J. geol. Soc. Lond.*, *69*, 223–51. (*Williamsonia spectabilis*)

174 Thomas, H. H., 1915. *Phil. Trans.*, *B 207*, 113–48. (*Williamsoniella*)

175 Thomas, H. H., 1925. *Phil. Trans.*, *B 213*, 299–363. (Caytoniaceae)

176 Thomas, H. H., 1933. *Phil. Trans.*, *B 222*, 193–265. (Mesozoic pteridosperms)

177 Thomas, H. H., 1947. *Adv. Sci.*, *4*, 243–54. (History of plant form)

178 Thomas, H. H., 1952. *Palaeobotanist*, *1*, 435–8. (*Glossopteris*)

179 Thomas, H. H., 1955. *Phytomorph.*, *5*, 177–85. (Mesozoic pteridosperms)

180 Thomas, H. H., 1958. *Bull. Br. Mus. nat. Hist.* (*Geol.*), *3*, 179–89. (*Lidgettonia*)

181 Thomas, H. H. and Bancroft, N., 1913. *Trans. Linn. Soc. Lond.* (*Bot.*), *8*, 155–204. (Cuticles of cycads and Bennettitales)

182 Thomas, H. H. and Harris, T. M., 1960. *Senckenbergiana*, *41*, 139–61. (Jurassic cycads)

183 Thompson, W. P., 1919. *Bot. Gaz.*, *68*, 451–9. (Phloem of *Gnetum*)

184 Townrow, J. A., 1956. *Avh. norske VidenskAkad. Oslo*, 1–28. (*Lepidopteris*)

185 Townrow, J. A., 1957. *Trans. geol. Soc. S. Afr.*, *60*, 21–56. (*Dicroidium*)

186 Townrow, J. A., 1960. *Palaeontology*, *3*, 333–61. (Peltaspermaceae)

187 Townrow, J. A., 1962. *Bull. Brit. Mus. nat. Hist.* (*Geol.*), *6*, 289–320. (*Pteruchus*)

188 Traverse, A., 1950. *Am. J. Bot.*, *37*, 318–25. (*Mesoxylon*)

189 Turrill, W. B., 1959, in *Vistas in Botany*, ed. W. B. Turrill. London. (Gymnospermae)

190 Vasil, V., 1959. *Phytomorph*, *9*, 167–215. (*Gnetum scandens*)

191 Vishnu-Mittre, 1953 and 1957. *Palaeobotanist*, *2*, 75–84, *6*, 31–46. (Pentoxylales)

192 Walton, J., 1926 and 1931. *Phil. Trans.*, *B 215*, 201–24, *B 219*, 347–79. (Carboniferous fern-like fronds)

193 Walton, J., 1949. *Trans. roy. Soc. Edinb.*, *51*, 719–28. (*Calathospermum*)

194 Walton, J., 1953. *Adv. Sci.*, *10*, 223–30. (Evolution of pteridosperm ovules)

195 Walton, J., 1957. *Trans. roy. Soc., Edinb.*, *63*, 333–40. (*Protopitys*)

196 Walton, J. and Wilson, J. A. R., 1932. *Proc. roy. Soc. Edinb.*, 52, 200–7. (*Vertebraria*)

197 Watterkeyn, L., 1954, 1959 and 1960. *Cellule*, *56*, 105–46, *60*, 7–78, *61*, 81–95. (*Gnetum africanum*)

198 Wesley, A., 1963, in *Advances in botanical research*, Vol. 1, ed. R. D. Preston. London and New York. (The status of some fossil plants)

199 Wilde, M. H., 1944. *Ann. Bot.*, *8*, 1–41. (A new interpretation of coniferous cones—Podocarpaceae)

200 Williamson, W. C., 1870. *Trans. Linn. Soc. Lond.* (*Bot.*), *26*, 663–74. (*Williamsonia gigas*)

200b Yamamoto, Y., 1932 *Icon. Plant. formos., Suppl.* 5, 7–11 (Amentotaxaceae)

201 Zimmermann, W., 1933. *Palaeobiologica*, 5, 321–48. (*Williamsoniella* with axillary flower-buds)

202 Zimmermann, W., 1952. *Palaeobotanist*, *1*, 456–70. (The main results of the Telome Theory)

ADDENDA 1971

203 Pant, D. D. and Nautiyal, D. D., 1967. *Phil. Trans.*, B 252, 27–48. (Fertile *Buriadia*)

204 Banks, H. P., 1970. *Biol. Rev.*, *45*, 451–4. (Geological time scale)

ADDENDA 1974

205 Beck, C. B., 1971. *Amer. J. Bot.*, *58*, 758–84. (Anatomy and morphology of *Archaeopteris*)

206 Bonamo, P. M. and Banks, H. P., 1967. *Amer. J. Bot.*, *54*, 755–68. (Fertile *Tetraxylopteris*)

207 Carluccio, L. M., Hueber, F. M. and Banks, H. P., 1966. *Amer. J. Bot.*, *53*, 719–30. (Anatomy and morphology of *Archaeopteris*)

208 Crepet, W. L., 1972. *Amer. J. Bot.*, *59*, 1048–56. (Pollination mechanisms in *Cycadeoidea*)

209 Kräusel, R. and Weyland, H., 1933. *Palaeontographica* *78*, 1–44. (*Protopteridium*, now called *Rellimia*)

210 Leclercq, S. and Bonamo, P. M., 1971. *Palaeontographica*, *136*, 83–114. (Fertile regions of *Rellimia*=*Milleria*=*Protopteridium*)

211 Leclercq, S. and Bonamo, P. M., 1973. *Taxon*, *22*, 435–7. (*Rellimia* – a new name for *Milleria*=*Protopteridium*)

212 Norstog, K., 1972. *Phytomorph.*, *22*, 125–30. (Archegonia of Cycads)

213 Phillips, T. L., Andrews, H. N. and Gensel, P. G., 1972 *Palaeontographica*, *139*, 47–71. (Two heterosporous species of *Archaeopteris*)

214 Scheckler, S. E. and Banks, H. P., 1971. *Amer. J. Bot.* *58*, 737–51. (*Triloboxylon*)

215 Scheckler, S. E. and Banks, H. P., 1971. *Amer. J. Bot.*, *58*, 874–84. (*Proteokalon*)

Index

Page numbers in *italic* refer to illustrations